## WHAT OTHERS ARE SAYING...

*The writing is excellent and the topic is intriguing. Hollis has a remarkably broad perspective on rapid product development in general and prototyping in particular. His knowledge of doing business in China is impressive.*

— **Terry Wohlers,** Industry Expert
Wohlers Associates

Better Be Running! *provides the basic understanding of how to do it right the first time rather than losing the time and spending the money to do it over. Imagination rules the world and if we accept the premise that we don't know what we don't know we will be able to continue to play a lead role in providing the innovator a solution to bring his products to market faster, cheaper and better. We must live every day knowing that whatever we are doing today is not good enough for tomorrow. If you need a widget you* **Better Be Running!**

— **Randy Barko,** Former President
Nypro Medical

*As a designer, I knew that each decision in the birth of an idea has a dramatic effect on the feel of the final product. Those decisions we make subconsciously are greatly enhanced by our ability to touch in a three dimensional world. 3-D printing is a fast way to get that early feedback and information into the creativity of the design. TOUCH IS THE MOTHER OF PRODUCT DESIGN. CREATE AN IDEA, EXPERIENCE THE PART, SEE A DRAWING... 3-D.*

— **Bill Masters,** Inventor
3D Printing with BPM

*In business and in life, I have been drawn to people that are passionate about what they do — especially when it involves manufacturing. Anyone can get lucky and succeed in business, but those that are truly passionate take the time to figure out a better way, share their experiences and enjoy every step of the process. Ron has combined years of knowledge and his unwavering commitment to make things more efficient to create a blueprint for those who want to be leaders in product development.*

*You can't teach passion — but this book allows you to feel it.*

— **Mitch Free,** Founder & CEO
MFG.com

*"It's about time for a book like this! America must stop ignoring the manufacturing elephant in the room and get with the program…"*

— **Ken Cooper,** NASA
Advanced Manufacturing Manager

*Until recently, innovation and product development was primarily for premium companies looking to maintain a premium position within their marketplace. Today, innovation and product development are just a starting point for any industry at any price point in order to survive. A company's future survival will be dependent on their ability to truly understand their customer base, dream up new solutions and implement them faster, accurately and better than their competition. It all starts with leaders using leadership techniques. Better Be Running! does an exceptional job of teaching these leadership techniques as well as enables companies to go from idea generation to market implementation at a faster more accurate rate. A must read for any company conducting product development.*

— **Russell Kohl,** President & CEO
Freud

*An engineering degree does little to provide one with the skills and tools needed to create products which stress design for manufacturability. Unfortunately, the outsourcing of design and manufacturing to countries such as China limits the job opportunities needed to develop our engineering workforce. My experience shows that China and Taiwan are very good at production, but reliance on them in the development/prototyping phase can greatly impact the timeline of a product. The real challenge is to be aware of China's tooling capabilities in order to capitalize on the production possibility frontier, while at the same time, striving to minimize cost and engineering change order cycles. Better Be Running! is a roadmap to guide the engineering and design community into the future of product development.*

— **Ryan Foss,** Mechanical Design
Engineer DeLorme

# BETTER BE RUNNING!

## TOOLS TO DRIVE DESIGN SUCCESS

BY RONALD L. HOLLIS, Ph.D., P.E.

**BETTER BE RUNNING!**
**Tools to Drive Design Success**
Copyright © 2007 by Ronald L. Hollis. All rights reserved.

ISBN-13: 978-0-9795760-2-7 (trade paperback)
LCCN: 2007927607

Published by:
CLSI
301 Perimeter Center North
Suite 500
Atlanta, GA 30346

*Info@betterberunning.com*
*http://www.betterberunning.com*

Unattributed quotations are by Ronald L. Hollis.

Other editions available:
Large Print 978-0-9795760-0-3
LIT 978-0-9795760-9-1
PDF 978-0-9795760-1-6
Mobipocket 978-0-9795760-8-3
Palm 978-0-9795760-7-5
Audio, MP3 978-0-9795760-6-7

Printed in United States of America

**Publisher's Cataloging-in-Publication**
*(Provided by Quality Books, Inc.)*

Hollis, Ronald L.
    Better be running! : tools to drive design success /
by Ronald L. Hollis.
    p. cm.
    Includes index.
    LCCN 2007927607
    ISBN-13: 978-0-9795760-2-7 (trade pbk.)
    ISBN-10: 0-9795760-2-4 (trade pbk.)
    ISBN-13: 978-0-9795760-0-3 (large print)
    ISBN-10: 0-9795760-0-8 (large print)
    [etc.]

    1. Production engineering. 2. Product management.
I. Title.
    TS170.H65 2007            658.5
                         QBI07-600151

# Table of Contents

# Dedication

*To every team member of Quickparts.com, past, present and future, for making us better every day!*

# Acknowledgement

W hat a great section of the book. It is like giving the acceptance speech for the Grammy's! You get to babble about all the folks you can remember that gave you the boost to succeed. With a book, it is easier in that you get to structure your thoughts and take as much time as you desire.

So, here it goes. Of course, it takes a team to achieve victory and have a positive influence on our environment. The creation of this book is no different, just the team is bigger or more talented. There were scores of people who, directly or indirectly, contributed to the development of this work. Each added their own special value to the process to make the best product we could to serve the reader.

While there were many that have contributed, the ones that have a special place in this book are the reviewers who helped improve the writing and presentation of the information. This team included Brian Ford, Roshan Darji, Daniel Ng, Bob Croteau, Cameron Moore, Collin Webb, Deepesh Misra, Ben Johnson, Randy Barko and Patrick Hunter. Also, much appreciation goes to the talented Art Siegert for the cover and illustrations.

We express acknowledgement to the competitors of Quickparts, for being so bold to copy what we do well and making it their own. We respect your success, appreciate the competitiveness and hope you are using this opportunity to develop your people to be positive contributors to the world. We all have a role to play in driving change. Particular recognition goes to **ZNMNJNH** and

BETTER BE RUNNING!

**OPLPJGHDU** (obviously a code word for the super intelligent in the crowd smarter than Robert Langdon) for their continuous and sincerest "flattery."

Special acknowledgement goes to those who were not directly involved in the book, but contributed over the years with personal relationships. These include Michael Maurice and Mark Mackie for having the confidence and courage to help me start a world-changing enterprise in 1999, when the future was fuzzy. Sameer Vachani and Patrick Hunter, who epitomize A-players at Quickparts. Greg Crabtree for being a consultant, investor, director and most importantly a friend the past several years as we have grown Quickparts and Dick Holloway for being the first angel to Quickparts.

Super special acknowledgement goes to Kim Brown, my assistant, for keeping my work life organized so I can do more with less.

As Jim Carrey says, "Behind every great man is a woman rolling her eyes." Probably true but the woman behind me has hid it well. I attribute all the success I have achieved to my GOD and my wife, Melanie. She is my wife, best friend, and mother of my son. She has always believed in my dreams and supported me to their success. She is probably the only person in the world who understands me well enough to put up with my crap.

There are two important people who deserve acknowledgement since they sowed the seeds that reaped this work. Virginia Hollis, my mother, for always being a provider to the family even when it was not easy to do so. The late Eddie Wicks, the person in my life that taught me how to be man and gave me the confidence at an early age to believe that I could become whatever I wanted in life.

It is my personal desire that the information in this book can help others make a positive change to the world. The most difficult part is in believing that you can!

# Foreword

There is a void in the product development marketplace for a book that educates business leaders about critical information on currently available technologies. With knowledge, these technologies can be implemented into the product development process to get products to market faster as well as reduce development budgets. The purpose of this book is to help fill that void.

Unfortunately, it is very common for engineers today to lack an understanding of how their virtual Computer-Aided Design (CAD) model actually becomes a real part through manufacturing. Many engineers don't know injection molding from blow molding. This kind of ignorance increases the chance that a company will spend millions of dollars on a product that cannot be manufactured, or worse, cost hundreds of times more than it should.

Many have heard of rapid prototyping, additive fabrication, or digital manufacturing technologies. Others know they are categories for SLA, FDM, and SLS. Most think they understand how urethane parts are cast and CNC parts are made. Several have or will have experience with injection molding, either for pre-production or production, and every developer today has to understand the world of China and how it fits in the manufacturing strategy. With the aid of this book, the guessing can stop, and everyone can **have the knowledge to properly and efficiently develop their products with these latest technologies and strategies.**

In using the latest technologies to quickly make parts, many engineers have experimented, but never fully understood what they were getting. Many will buy a Stereolithography (SL) part without realizing it is only one strategy in a plethora of solutions. One technology cannot and does not answer all of your problems. Most small service providers specialize in only one manufacturing technology due to the cost of the equipment and the specialized training required. They will sell you only what they have and know, not what you need. It is important to remember that **the purpose of the part should determine the manufacturing technology to be used.**

This book focuses on manufacturing processes, tooling choices, and winning production strategies. My goal is to share their strengths, limitations, benefits, and best applications. Beyond that, I intend to show the relationship of the manufactured part to product development and how product development thrives in free enterprise systems, feeding business, community, governments, and global partners in an ever-expanding spiral.

My passion is building technology-based businesses. With this same passion, I have invested time and energy into this book. Its purpose is to "teach and delight" the professional, CAD-proficient engineering community about the latest and least understood product development tools. If you are an engineer, manager, or executive in any industry's product development field, this book's "street-wise" information will provide you with the foundation to make great decisions in your product development process. You will learn to turn your virtual world into reality, quickly! You will learn how to achieve solutions faster by reducing inefficiencies and unnecessary steps in the transition of your design from the virtual world of CAD to the real world of manufacturing.

Another part of my mission is to educate engineers to become better product developers. We must ask ourselves: How can I be great? In design, one fundamental of being great is knowing how to verify that the virtual will perform in the reality. The technologies discussed provide the options for verification strategies. With each new application that confronts you, you will have to pick the best strategy. Just as there are many ways to go home after work, such as by bus or car or foot, you have to pick the best product development strategy based on your resources.

If you are looking for a highly technical textbook with details of prototyping processes, descriptions of lasers, and formulations of plastic material, you have the wrong book. There are plenty of great "geek" books on the market to tell you more than you could ever want to know about the technicalities. This is a simple, practical, user-friendly guide that explains how to apply new technologies to make a noticeable difference in your design process. By becoming an informed consumer of these processes, you will be able to save time and money, oftentimes 50% or more, a level considered heroic in most companies. Equally important, this book shows you how to prevent wasting time and money in your product development process. At Quickparts, we are focused on helping our customers make their product development process more efficient. We do this by applying proven techniques and technologies to maintain a reliable outcome. (We leave research and development to universities and other companies that are not focused!) At Quickparts, we aim to be directly or indirectly involved in the procurement of **every custom-designed part in the world.**

The world of manufacturing is always evolving. A whole new world of layered manufacturing and low-volume production is now at your fingertips. To help convey the practical aspects of these technologies, we are using a manufacturing fairytale to illustrate their applications. To entertain our readers and display the dynamic educational spirit of Quickparts, we have included cameo appearances of a fictional product development superhero, Johnny Quickparts. Johnny is a quiet, shy geek who inadvertently becomes successful through his earnest love of knowledge. If the bizarre challenges of Johnny and Acme Design Corporation typify your work day, then you are normal! We hope that Johnny's heroic achievements and romantic forays make this hefty plate of technical limitations not only palatable but also enjoyable.

Radical advances in 3-D modeling software and equipment in the '80s and early '90s have super-accelerated product cycle times. If professionals aren't yet using these tools, it's only because they don't know about them or don't fully understand how to use them. If you read and retain all the information herein, you will be more advanced than most professionals in product development today!

This book was designed to give you a basic foundation in the new paradigm of manufacturing. Our intention is that the reader will absorb enough solid information here to be able to seek out more specific information as required. You may even become conversant enough with these technologies to impress strangers at your wife's company Christmas party! Not many companies in the world would share insider information with you, but Quickparts is different. **We are in the business of truly serving our customers,** which is a core value of the company, and drives all of our decisions, including the decision to invest time and money to create this book.

Finally and most importantly, we hope to ignite innovation in our readers with an insight into the future of "global product development." Leveraging the strengths of China and North America will drive the efficiencies of product development to the manufacturing sector. For the US, the way to economic victory is to drive innovation continuously in new products, to levels that no one else in the world can match. At Quickparts, this is our dream, our goal, and at the end of the day, our reality.

*Ronald L. Hollis, Ph.D., P.E.*
*President and CEO*
*Quickparts.com, Inc.*
*Atlanta, Georgia*

*May, 2007*

# Chapter 1

# Everything Matters

## THE BUSINESS OF PRODUCT DEVELOPMENT TECHNOLOGIES

*"Every morning in Africa, a gazelle wakes up. It knows it must run faster than the fastest lion or be killed. Every morning a lion wakes up. It knows it must outrun the slowest gazelle or starve to death. It doesn't matter whether you are a lion or a gazelle... When the sun comes up, you'd* BETTER BE RUNNING!*"*

## ⁞⁞ Product Development—The Impulse to Create

While the gestation time of a human fetus has remained the same for millions of years, the time to get your product to market has accelerated a hundredfold in only 20 years. We imagine ideas and "birth" products at a super-accelerated rate today. While the desired outcome is the completed product ready for your customer, we as product developers still need to ensure that we are producing what we really intend to produce. The design must reflect the intent of the functional part.

The high-tech processes and tooling strategies addressed in this book represent a radically new way to get a product on the shelf as quickly and economically as possible. The latest innovations of Computer-Aided Design (CAD) and Computer-Aided Manufacturing (CAM) technologies have significantly improved product development speeds, especially when combined with the continuous innovations of rapid technology tools. This new "toolbox" full of little-understood options is critical to quickly birthing ideas into real products faster, easier, and better than ever before. In learning this quantum approach, engineers and manufacturers need to see the whole before focusing on the part.

In discussing the complexities of product development in the twenty-first century, with recent manufacturing advances and new global players, a context showing macrocosm and microcosm is useful. In today's global society, interdependence has replaced isolationism. The world has opened up well beyond our own company, city, state, and country. It's now considered de rigueur for companies to spend a great deal of time overseas, while in the past, it wasn't even in our imagination to do so. In other words, product developers now have more freedom to shop for each element of manufacturing from a "global superstore." However, learning to make the right choices in a suddenly smaller world can be frustrating and costly. Learning anything complex is a challenge and requires that the subject matter be broken into its most simple elements.

With that said, this book describes the best way to manufacture in the twenty-first century using new processes that super-accelerate product development times. The subject matter is divided into three main areas of expertise: high-technology part building, low-technology manufacturing, and tooling strategies for aggressive product development.

Parts really are the center of the universe, and if you don't understand or agree with this at the moment, you will by the end of this book. Dramatic cost and time savings are powerful motivators to learn all the tools in the rapid technology toolbox as well as the shape-shifting technologies related to CAD and CAM processes. In very little time, you will have every bit of knowledge you need to become your company's hero.

## ⠃⠃ The Power of Product Development in Free Enterprise

*"製作的零件是宇宙的中心 is Mandarin for*
*manufactured parts are the center of the universe!"*

Product development is the oxygen of business. Free enterprise nurtures the creativity behind product development and unleashes powerful, positive change in the world. In the last 25 years, product development has "gone global" and continues to race after an ever-widening horizon, be it the far field of innovation or the geopolitical turf of Southeast Asia. Product development flourishes in an environment of freedom. The business future looks bright in countries where businesses thrive as the primary elements of free enterprise, like China's recent firestorm of economic expansion. Manufactured parts, derived from businesses, are the common denominator to both free and un-free societies. **The almighty part is the sole reason the US now connects to the previously inaccessible society of China.**

If you study the incredible "power of the part," you can see how it feeds into an ever-expanding spiral of interconnection. Follow the movement of each element within manufacturing and discover the ripple effect set in motion by the part. Parts generate revenue to sustain business, business sustains employees, employees sustain communities, communities sustain governments, and governments sustain other governments that, we hope, sustain a peaceful world in which we are all interdependent. This is why, in every language, parts really are the center of the universe.

A potent driver of society, product development businesses in a laissez-faire economy are required to develop products to grow more commerce. The most dynamic product development teams require dreamers, doers, innovators, and leaders who continually add to their knowledge base, as the most far-reaching knowledge drives the greatest innovation. **The latest**

**business mantra "faster, better, cheaper"** is an unrelenting standard to which we all answer. It impacts the way we imagine a new product, the way we conceive a well-designed assembly, and the way we verify, test, and produce tools and parts to make the final assembled product at the greatest quality and the least cost.

Related to business freedoms, the United States has the greatest political freedom of any country on earth. As individuals and business leaders, we each hold this freedom in our hearts and hands. Unrestricted freedom as a social value can create any possibility and find the solution inherent in any problem. As a country that consists of many great businesses—the offspring of freedom—we Americans have somehow lost our power to control destiny, as evidenced by the economic doom and gloom reports. The manufacturing sector has incorrectly deduced that if it can't manufacture products as it had in the past, then there is no economic hope. While economic hits have been severe due to offshore movement, the US product developer, engineer, and manufacturer can overcome much of the drain if they begin to work closely together and add value to the product development process. As a country and a global leader, the US must grab hold and steer product innovation, for **whomever rules innovation has the power.**

The American freedom to develop products is an important counterbalance to offset the impact of China's growth and power, which many countries are still learning to accommodate. While much manufacturing has indeed moved out of the US, no one can own innovation, the most beautiful freedom of all. Innovation is like the mythological Siren that you can hear but can't see. Her song is always beckoning you toward horizons undefined.

Now is the time to take off our business blinders and invoke creative options for a new manufacturing era. The US is blessed with a government that ensures freedoms, and therefore a positive future. What we have now is far better than the alternative: no freedom, no future. The world of today requires continuous innovation to maintain the status quo of livelihood. What we have now, the potential for innovation, can catapult product development to new heights never before imagined. This is a time when heroes emerge.

## ⠿ It's a Global Neighborhood After All

*"Put thirty spokes together and call it a wheel;*
*but it is in the space where there is nothing*
*that the usefulness of the wheel depends."*
—FROM THE ANCIENT CHINESE
PHILOSOPHICAL TEXT, TAO TE CHING

The word global doesn't mean what it meant 25 years ago. Global isn't global anymore. We have new neighbors whose mind-set we are desperately trying to understand. While the American business mind has been shaped by over 200 years of unfettered power to choose and create, China has exercised freedom for only 35 years. Moreover, American product developers live in a strange time when we are now doing business with and, in fact, relying on this long-perceived adversary.

From an American point of view, China's current business environment is difficult at best, for it seems open and closed, free yet restricted. China offers traders from afar a paradoxical system, a unique hybrid of communism and capitalism. The US is now intricately interconnected to China for one reason and one reason only: to make parts.

Like it or not, we are all one. The exciting news is that local communities expand exponentially into global communities. Realizing the connection between everything and everyone inspires us to be better servants to the world, our collective home, by doing the best we can every day. In life and business, everything matters...even the butterfly.

## ⠿ Freedom to Innovate and Develop Products

Long before a cash economy took hold, business began as barter. Neighbors traded fish to get corn. A good rifle was worth one mule. Back then, no one could imagine a Starbucks or a four-dollar cup of coffee. No one could envision a future where people would blatantly use a coffee bar as free office space, complete with wireless connections and spirited entrepreneurs shouting private business details into their favored communication device. Out of freedom, we have created so much good and so much weirdness, like the indoor sundial.

As an American product developer, it doesn't matter whether you are in Los Angeles or Shanghai; your innate ability to innovate new product is independent of governments. You have the power of choice, the freedom to imagine. While consumers in society will dictate the success or failure of your product, the American product developer is engrained with the freedom to think and create. This unrestricted freedom to create fuels a product development business, which exists solely to create products. In China, where freedom of expression has been stifled for many years, a dearth of creativity reveals itself in functional but mediocre products.

New product development is inherently related to our personal pursuit of happiness and personal freedoms. Economist Adam Smith, born in 1723 and the author of *The Wealth of Nations,* created a doctrine of free enterprise that is the cornerstone of our capitalist market. Thanks to our spirited forefathers who protected the US with economic freedoms, product development thrives today. For the past 100 years, their guiding commitment to defend freedom in all parts of the world has fostered an environment in which we as citizens can freely pursue "life, liberty, and the pursuit of happiness." In other words, our visionary, inventive ancestors created the US as an environment of freedom. In today's world of laissez-faire economics, we can define and launch freedom in whatever fashion we choose. We live in a system of capitalism where you, as the customer, get to vote every day on the future existence of every business. Walking into a store, you are "voting with your feet." As a consumer or business leader, the impact of your freedom is profound.

In the US, we often take our exhilarating freedom for granted. A capitalist from the US has unlimited freedoms to develop products compared to a communist capitalist from China with very limited freedoms. In the twenty-first century, the new Chinese system of "capitalistic Marxism" is like the unusual ugli fruit, believed to be a chance hybrid between a mandarin orange and grapefruit. If we realize that free enterprise is comprised of entities called businesses, we can begin to understand how businesses, governments, and the market are all interrelated, "ugli" and beautiful.

**QuickTip:** China believes it is destined for global domination. Americans have to decide the role they will play in this new world. Innovators, step forward!

## ▦ A Great Threat, a Great Opportunity—China

Some 200 years ago, emperor and military leader Napoleon Bonaparte prophetically commented, "Let China sleep, for when she wakes, she will shake the world." The new manufacturing capital of the world, China, is officially awake and she touches everything you buy, like it or not. But the same great leader also said that "Money has no motherland," and "Imagination rules the world." The economic pie isn't shrinking, it's growing, and it's there for the innovator to seize.

China's new freedom presents a high-level conundrum. Business in China is very free yet very restricted under a Marxist government, proof positive that freedom is a potent force. As the Chinese adapted new freedoms over the last 35 years—freedoms that drive real power—the country expanded quickly as the manufacturing center of the world. Like a two-headed dragon with a split personality, the Chinese system of capitalist communism is confusing at best but workable with the capitalist system of the US. China seems able to allow sufficient economic freedom for businesses to thrive while maintaining tight political control over other aspects of life.

The US workforce can learn a great deal by studying the zealous Chinese work ethic. With so much new freedom, China is like the US of the early 1900s. China businesses work much harder and express more entrepreneurial spirit than the US of the twenty-first century. Supplementing its passion to succeed, China has had access to our universities and has greatly benefited from learning from our well-documented mistakes as well as successes. This access to knowledge has radically driven China's economic expansion unlike any other country in the history of the world. In contrast to China, the USSR resisted learning, changing, and free enterprise until its fall. While the pressures of the US during the Cold War initiated its demise, the USSR's lack of economic expansion from a free market dealt the final death blow.

In this global context of threat and opportunity, the US must maintain a leadership position by driving innovation and leveraging the world as a product development resource, using the best of all places and people. The American product developer lives in a time when business is a unique and powerful vehicle that can have the most significant impact on the world. In a free-enterprise system, business can provide livelihood and happiness to individuals, produce leaders, develop communities, and build a future for countries. Within a dynamic structure of capitalist business, we can drive out waste and require that markets operate efficiently at optimal performance. Therefore, product developers must focus on the core competencies of each region of the world to develop more products "faster, better, easier."

China has had ample time to ramp up, and now poses not only a great threat but also a great opportunity. For manufacturers feeling defeated by China's super-success, a few final watchwords from Napoleon: "You become strong by defying defeat and by turning loss into gain and failure to success."

## ▓ The Value of Innovative Technologies

> *"Products that get to market fulfill consumer
> needs and wants, which increase the pleasures
> of life, which then drives the happiness of
> the world, and further frees the creativity of
> mankind. And you thought it was just a part."*

### *Design on Monday, Manufacture on Wednesday*

"Dream it, Do it!" Today's product development cycle is almost
that fast, thanks to an abundance of new product development
tools. You can design a product in Atlanta on Monday, get your
investor's blessing in New York on Tuesday, and manufacture
it on Wednesday—an almost unthinkably fast creation story.
When the product developer of the twenty-first century realizes that the world really is his or her toolbox and understands
how to apply these new tools, nothing can stop their glorious
success. Isn't this a strong enough motivator to learn a **new
manufacturing paradigm?**

Some 30 years ago, American product design became compla-
cent, appearing lackluster in innovation. The illusion of product
development prowess in the US was still strong. In reality, it was
a time of exploding Ford Pintos, heavy telephones, and clunky
televisions. Fat, dumb, and happy, design companies were sleep-
walking, and product developers sat in a trance and seemed to
be resting on collective laurels of the US as a "success culture"
of the '50s and '60s.

In the '70s, global competition reared its head for the first
time. Fortunately for the US, Japan decided to intervene and
take over the US market in everything. The old corporate
geezers in the US did not want the world to change until they
retired. The good ol' boys denied the takeover and stuck their
heads in the sand while counting the days to retirement. These
long-timers shifted the problem to the next generation, letting

Japan's elegant know-how be someone else's problem. Meanwhile, Japan was buying up all the real estate in the US and gaining product market share.

> **QuickTip:** A dynamic cultural renaissance of innovation requires that we are always changing our definition of the present. Never accept the status quo.

As it turned out, the Japanese invasion of the US market shook things up and released a new wave of innovation and corporate development in the '80s. Out of obvious necessity to compete globally for the first time, the next generation of Americans germinated many innovations, including CAD software and sophisticated technologies that convert the output of the CAD software to real parts using additive fabrication (AF). Additive fabrication was called rapid prototyping (RP) in the beginning and is still a common term used.

In the '90s, innovation was revealed through more sophisticated 3D software and newly released additive fabrication machines. The turn of the century gave the mass market access to that innovation, now at the point of fully evolving into new realities of layered manufacturing and low-volume production.

Some 20 years ago, US companies began sending their manufacturing orders, for the first time, to Asian countries to maximize profits. Prior to this, offshore manufacturing had never dawned on anyone until giants like IBM realized that they didn't have to make what they sold. Many companies followed their footsteps and manufactured in Taiwan, Singapore, and China, forging a "smaller," more interconnected world. The fear mongering of the American manufacturer was that our future was being exported for the sake of near-term profits, a notion that was mostly correct. The offshore trend completely reorganized the manufacturing world. Now, we all have to adapt to the reality of a smaller world with global neighbors peering over the fence.

Unfortunately for the US, very few manufacturing businesses proactively accepted responsibility for seeking efficient ways to reposition themselves in a smaller world. Instead, most companies sought entitlements such as tariffs and government protections. Many even tried to prevent change from happening. This time of turmoil uncovered a blind spot in product development processes: a thick but invisible wall had always divided manufacturing and engineering. Throughout the '90s, the division seriously affected decisions of both parties, but we didn't yet fully realize the wall was there and it was bad for everyone. When all the manufacturing was moving to China, engineers were not crying about it. They did not yet realize that "No engineer is an island." Back then, no one could see the dangers inherent in keeping the two disciplines separate.

Lessons learned show that companies cannot effectively separate manufacturing from engineering, even though they can build "virtual walls" between them. Like husband and wife, they still need to work together, communicate, coordinate, and collocate—at least in the same continent! Somehow, the reigning brains in highly compartmentalized corporations had overlooked what was really happening in their product development process. They hadn't noticed the engineer when he was on the factory floor helping to solve a production problem. They hadn't heard the decisive conversation when the tooling manager was in the engineer's office telling him to make a change so that the part would actually be manufacturable. It was this almost intangible, non-measured, non-monitored communication between departments that was the "glue" in great product development. We didn't know this until the two disciplines were separated by thousands of miles.

> **QuickTip:** Engineering and manufacturing are now in a symbiotic marriage, and their existence depends on each other.

Product development folks, not in the manufacturing arena, also rationalized that the issues with manufacturers were isolated to manufacturing and not the problem of the educated engineer. The engineers and designers were "protected," since they had always been the brain trust with knowledge, intellect, and education. The world would always need these smart development professionals from the US to develop new products, right? We had to learn the hard way that this was not true. It was bad for these marriage partners to exist half a world away. The realization that development professionals needed to be near manufacturing affected many lives and livelihoods. As many product development issues were coming out of Asia, engineers from US companies were required to be in China.

A subtle trend is now emerging as engineering opportunities continue to subside in the US. China is expanding its role into product development by "owning" more manufacturing and more engineering. Today, American companies see less design work and do more project management as part of this trend. As US product developers find themselves no longer designing or making the product, they know the challenge is to become smart enough to maintain control of the product development process.

**QuickTip:** Innovation is the final competitive advantage.

The solution to our competitive challenge is to drive innovation continuously in new products to levels that no one else in the world can match. While this is much more complicated than just making our own parts or designing the product and shipping the data to China for manufacturing, heightened innovation is our last option to maintain our economic freedom, drive our economy, and ensure the future freedoms of our country.

### Drive Innovation

Being an innovator is not trivial. Product developers have to promote a dynamic cultural renaissance of continuous change. Our perspective should be one of looking through a lens at a much bigger picture in an almost detached, allowing, and inviting way, without ever getting stuck or side-tracked. Something truly dynamic is always changing in present time. Evolution rarely stops or has a well-defined current state. To get your mind into a bigger perspective, remember that in 1990, none of us visualized any of today's fabulous products. We didn't know and couldn't know what was coming or what was possible. What if a well-meaning bean counter had squashed a dreamer's iPod vision? **Innovation is meant to be nurtured, not dismissed!**

If American product development succumbs to complacency, innovation will spiral downward rapidly. The old days of prowess in product development will be a faint mark on a timeline of US history. To prevent this, we need to ask ourselves on a daily basis: Where are we, and how did we get here? The answer is a strong responsibility pill: **Performance is the consequence of a series of decisions.** In every company and country's history, many decisions are made by many people. The current performance of our country—and our companies—is the consequence of all of these decisions.

Most American manufacturers tend to lose their bearings in a global marketplace and wear a "V" (for Victim) on their chests. Remember that the US is home to a huge number of entrepreneurs and innovators. As a culture, we need to take heart and rededicate ourselves to continuous learning and innovation. If the US manufacturer or product developer plays the victim role in a global society, its customers will take their business to China and have it designed, tested, analyzed, and manufactured there, for themselves. US manufacturers must find the entrepreneurial spirit and own the words "faster, better, easier." We must embrace global competition in manufacturing and reposition ourselves for success in the US market. Just remember: Leverage the good of China and other parts of the world, and then focus on your core competency. **Accept reality as it is, not as it was!**

What can US manufacturers do to triumph over complacency? Every manufacturing business in the US has essential skills that every product development company needs to access. You can provide a great deal of value by making products better-designed for manufacturing, reducing costs and time, and leveraging the cost savings of global manufacturers for some phase of the process, such as tooling. By shifting business emphasis to adding value to innovation and design, you can still be very successful in the US. A manufacturer in the US can most definitely grow its business even though everyone else is running to China.

## Make the Investment

Product businesses should recognize their vested interest in free markets and economic freedom because a free market is about the exchange of products and services. The more products your company can develop, the more consumers will buy them. All you have to do is figure out how to do it faster, cheaper, and better. "Faster" is essential since developing more products generates more profits for your business; "cheaper," since strong competitive forces always exist among popular products; and "better," since your product needs to be truly remarkable as a differentiator. In essence, **you must be smarter, better informed, and more innovative to win.**

Finally, learning new technologies is critical in a fiercely competitive global market. You are responsible for invoking new options and widening your world. Whatever it takes to ignite your vision: jump outside the box, turn your thinking upside down, take a wellness day. Write down all those "crazy ideas" that would never work in a million years, or would they? Are you willing to make the investment?

## ⠿ Amazing Technologies and Strategies

The process of using advanced technologies for product development begins with the transformation of electronic representations (CAD files) into the physical world of real parts. The Product Developer's Toolbox is a choice selection of CAD-friendly product development tools that bring about product innovation faster and better. The Product Developer's Toolbox is defined as the application of technology and processes for the manufacturing of functional parts quickly and economically. It includes a mouthful of hard-to-say processes including: the AF processes of Stereolithography (SL), Fused Deposition Modeling (FDM), and Selective Laser Sintering (SLS). Low-tech manufacturing methods of Cast Urethane (CU) and Computer Numerically Controlled (CNC) parts are also included. The toolbox also includes tooling strategies that feature the ability to make plastic injection-molded parts from thermoplastic materials in days, and production injection molds in weeks. By using the Product Developer's Toolbox fully, you will eliminate inefficiencies in proofing product concepts, leaving you plenty of time to focus on innovation itself. After all, blazing hot innovation is the only way for the US to maintain its champion status in a global market.

### Tools in the Virtual Toolbox

Think of this book as a virtual toolbox of quick technologies with three main tools: high-tech part building, low-tech manufacturing, and tooling strategies for the twenty-first century.

### • High-Tech Part Building

In case you don't know, additive fabrication is a relatively new manufacturing process that began in the '80s. This new process added another dimension to the manufacturing world by complementing subtractive and formative manufacturing processes. AF was created by the development of a class of automated machine technology that quickly fabricates physical 3D parts from electronic 3D data by building the part in layers. Over the past 20 years, there have been many terms used to represent the process or the output of these technologies. Here we are using the term AF to represent the manufacturing process to make a part and then may reference to applications of AF for more specific descriptions, such as RP when we use AF to make a prototype or low-volume layered manufacturing (LVLM) when we use AF to make parts that are being used in production.

This book concentrates on the AF processes of Stereolithography (SL), Selective Laser Sintering (SLS), and Fused Deposition Modeling (FDM). Based on the market information since 2000, Stereolithography represents approximately 75% or the lion's share of the AF market, FDM is roughly 13% and SLS represents the remaining 12%. There are many other processes that exist in the AF market, but none have gained an appreciable percentage of the market. It is important to differentiate the AF processes from the up and coming 3D printers that are growing very rapidly. When these technologies are combined together, the statistics shift drastically.

Used in every major industry, AF offers many benefits, including the transition of a design concept to a physical prototype, and testing of form, fit, and function. Other benefits of AF include reduced lead times to produce prototyped components; improved ability to visualize the part geometry with a physical replica; earlier detection and reduction of design errors; increased capability to compute mass properties; and advantages in the elimination of waste and costly design changes.

Reminiscent of the imaginary worlds of *Star Trek*, *Star Wars*, and *The Jetsons*, high-tech part building processes are nothing short of cool, fabulous, and amazing. The most advanced and dominant process, SL, commands most of the market for all parts produced from standard AF technologies. As the first and most popular liquid-based, laser-activated additive fabrication system, SL produces plastic parts layer by layer from electronic CAD data. SLS is another additive fabrication process used to create prototypes and functional parts. This additive manufacturing method creates solid 3D objects by fusing or sintering particles of powdered material with a hot $CO_2$ laser. FDM is the third additive fabrication process, considered the strongest, but slowest, of the major solid-based additive fabrication systems that also produce plastic parts from electronic CAD models. This equipment is characterized by a heated head with two extrusion nozzles that build a part layer by layer from extruded filament.

Rapid prototypes clearly illustrate product characteristics. Therefore, applications are many, including: customer presentations and demonstrations; proposal support; packaging studies; marketing studies; and review and collaborative discussions. Rapid prototypes also serve as master patterns for room temperature vulcanization (RTV) tooling and investment casting, as well as tooling for injection molding.

> **QuickTip:** The engineer converts inspiration into practicality.

## • Low-Tech Manufacturing
Other "low-tech" industry basics are included in the Product Developer's Toolbox, such as CU part making and CNC machining. While these processes are not new or rapid, CU and CNC have been the most reliable workhorses of the parts industry for decades, and are still essential to manufacturing.

## • Tooling Strategies for the Twenty-First Century
Advanced tooling strategies that support aggressive product development are the last tool in the virtual toolbox. To meet your goals, you must make a final decision on tooling choices for production, the most expensive phase of your product development cycle. Low-Volume Injection Molding (LVIM) is best for short runs up to 50,000 parts, in contrast to heavy duty production tooling that can produce millions. You can also use a combination of both types of tooling to beat your competitors with a very clever product development strategy. Keys to tooling, fully explained in the last part of the book, will ensure success for the riskiest part of the product development timeline: production.

## Evolution of Product Development Technologies
The purpose of RP is to get expert input on your design early in the product development process to reduce failures that are likely to

happen. Therefore, it makes no sense for today's product developer to skip the RP process. In the '80s, a designer made 2D drawings and gave them to a model maker who hand carved mostly prismatic objects. Design changes and extensive labor meant that getting approval on the design could take weeks or even months.

Product development radically changed in the late '80s when the virtual world involved high-end, high-cost 3D modeling software. Advanced technology germinated then pollinated the rest of the world. First used in very few specialized, high-tech programs, such as the Space Station, this technology has since proliferated to the masses, causing the paradigm shift in product development.

Using technologies significantly reduces errors and issues, thus saving time and money. Your products get to market faster, which increases your time in the market and generates more revenue from products sold. If the next widget you are designing will generate $1,000,000 per week in revenue, it doesn't take long to realize that the three weeks it would take to make an engineering change can cost your company millions in revenue and profits. Always remember that it is **profit that continues to pay your salary.** For that reason, we all have a vested interest in saving our companies money and time.

In the early '90s, the evolution of solid modeling in CAD made possible the electronic representation of 3D a reality. Late in the '90s, this software proliferated down into the middle market. Ten years later, 3D modeling proliferated into even smaller markets so that businesses with one or two people could design with it. Today, sophisticated 3D solid modeling software costs very little, which means that everyone should design in 3D and 2D drawings should be for reference only.

Now that every design is virtual, physical verification is required. Rapid prototyping, in reference to being a product development technology, is an extension of the modeling and verification process. If you are in the product development world, you already know it's an expensive arena. Every change that you don't have to make in tooling is a cost savings. Prototyping actually turns out to be free if you consider that, as an iterative design process, it validates your design prior to producing the tooling, launching marketing programs, and even mass-producing the parts.

Prototyping "insurance" verifies that what you think you see and imagine is what the part really is, once it comes into physical form. Engineers fear that they will lose time if they take the time to make a prototype. However, if you skip this critical step to save time, you invite more changes further down the road, which eats up even more time than you were trying to save. Correct any misperception that prototyping your design will cost you extra time and money. In fact, you save significantly on both if you invest in the prototyping at the front end of your design.

In the late '90s, it was common for a designer, engineer, or manager to spend a quarter of his or her time in sourcing and buying the parts needed to verify the product. Today, with online instant quoting, you buy custom-made parts as easily as buying books from Amazon.com, which is really fast! This leaves the remaining 75% of your time for administration, meetings, and actual development of the product. Using these new rapid technologies, you can now recover and invest almost all of that first 25% of your time in the actual product development process.

## ⠿ The Keys to a Brand New Lamborghini

*"Parts are like gas stations and parking spots; you only care about them when you need them."*

By now you know that knowledge is a vehicle. The information presented here is like the hottest car in the world. Once you get your mind around it, this technology can transport you into powerful and prestigious realms of product development. You've had time to think about the almighty part, and you've watched as the "power of the part" influences the expanding spiral of our interconnected life on this increasingly small planet.

As product makers, we are clearly not in Kansas anymore. This exciting new information will take you on an extraordinary road trip across an expansive paradigm shift in manufacturing. If you are not well-versed in CAD technologies, stop reading now and run to the nearest software provider to buy the latest version of SolidWorks, Autodesk Inventor, or Pro/Engineer. Get your hands on software that will allow you to evolve your process into the twenty-first century.

Just as you must learn to walk before you can run, you must learn each technical process before you can access the best product development solutions. In this book, we are running fast. Don't worry, though, our super-geek engineering hero (an insecure romantic), Johnny Quickparts from the lackluster Acme Design Corporation, will be running right by your side, showing you the how's and why's with practical, memorable applications.

Finally, the highest level of product development requires dynamic dreamers and doers. As you keep turning pages, ask the question, **"Where does innovation reside?"** What sparks in us those precious qualities of curiosity, creativity, and originality—all the elements of innovation—to just suddenly appear? If you hear an answer, shout it from the rooftops. The source of innovation will breathe new life into our world.

As Johnny would say,

**"Start your engines!"**

# Chapter 2

# Stereolithography

## THE LION'S SHARE OF RAPID PROTOTYPING

*"Rapid prototyping makes heroes."*

## *QuickSMART*

**D**efinition: Stereolithography (SL) means to print in three dimensions (stereo: three-dimensional; lithography: to print). SL is the first and most popular liquid-based AF system that produces plastic parts from cross-sectioned CAD data. Electronic CAD design data is converted to an STL (Standard Tessellation Language) file format. Special software slices the CAD model into thin layers and creates build instructions for the machine. Layer by layer, the Stereolithography Apparatus (SLA) machine replicates a plastic physical model out of photo-curable resin. The resin turns into hard plastic wherever touched by an ultraviolet (UV) laser.

**Why You Need It:** To reduce design cycle time by 50%; to ensure part functionality; to eliminate design changes late in the manufacturing cycle; to quickly create a single physical model or family of parts to touch and feel; to see how the part interacts with its environment; and to check physically/geometrically because drawing interpretation is prone to human error.

**Ideal Uses:** Trade show models; CAD verification; proof-of-concept; conceptualization; visual aids for marketing and planning; form, fit, and function testing; flow analysis, stress analysis, mock-up for testing, clearance checking, patterns for tooling and casting; tooling production, reproducing snap fits, assisting collaborative design, engineering, and manufacturing team in planning and decision-making; and a powerful communicator that provides complete information and understanding to all parties so they don't have to rely on guesswork or CAD data.

## ⁙ Stereolithography Background

As the forerunner of the latest Industrial Revolution which began in the early '90s, SL allows you to create a 3D plastic object from a CAD model in several hours. Prior to this technology, conventional prototyping methods could take days or even weeks. Whether you are a design engineer wanting to verify your concept, or a manufacturing engineer needing form, fit, and function feedback, SL gives you and your team a quick, accurate way to convert virtual data into real objects. It allows you to test designs in their physical environment before committing to expensive tooling.

If you are new to the exciting world of AF, which includes rapid prototyping (RP), rapid tooling, and low-volume production manufacturing, you have a strong cost incentive for remembering all the acronyms associated with these revolutionary processes. Lucky for you, there really are only three AF processes we are including in the Product Developer's Toolbox. Of course there are dozens of other processes that exist with varying degrees of utility to product development. The first we are discussing is SL, considered the pioneer of the AF industry.

Stereolithography was the watershed in manufacturing. It was invented by Charles Hull and made commercially available by 3D Systems, Inc., in 1988. Because 3D Systems was the first to market the SLA machine, many folks frequently misuse the term SLA to generically describe all RP techniques and any liquid-based, UV AF process. Within the industry, SLA has become as widely misused as the name Kleenex. One example of a process that produces parts that are similar to SL is made by Objet Geometries, Ltd. Objet machines produce parts using a different technique known as PolyJet. This technique jets or sprays a photopolymer resin instead of using a vat of the resin and solidifies this resin with a UV bulb instead of a laser beam. However, the parts are similar to parts made by SL, but have much smoother surfaces. Since 1988, over 40 AF systems have entered the worldwide market, competing to serve product designers, tool manufacturers, manufacturing engineers, and ultimately, the end consumer.

An informed customer knows that an SL part's strength, accuracy, and surface finish depend on variables of layer thickness, materials, and post-processing. Other parameters influence the performance and functionality of the parts, including the physical and chemical properties of the resin; resolution of the optical scanning system; laser type, power, wavelength, and spot size; the recoating system; and the post-curing process.

When using service providers, it's important to remember the variables, as the production of your SL part can be more of an art than a science. You need to thoroughly understand the process and parameters before cutting a purchase order to the service provider. **Many times customers don't know what they don't know.** They end up getting exactly what they asked for, which is not at all what they really wanted. Once informed, you will realize the benefits of these relatively new technologies that dramatically lower your product development costs and reduce your time to market.

## ▓▓ SL Process—Inside the "Replicator"

Design engineers jokingly refer to the SL process as *Star Trek's* "Replicator"—a machine that converts energy into matter— producing spare parts quickly to avoid starship disasters. A very apt comparison, SL converts virtual models to reality in short order, saving your company considerable budget and time. This laser-based process produces plastic parts by curing photo-curable resin with a UV laser system. The SL process is classified as AF due to the process of producing a physical part with successive layers. The SLA system consists of a UV laser, a vat of photo-curable liquid resin, and a controlling system. Your CAD data provides cross-sectioned build information to the SLA system. Layering technology is performed by computer software that slices the CAD data into layers, called slices, and outputs the slice data to the SLA.

### *Step by Step with SL*

First, an operator loads the STL file from your CAD data into proprietary software, which digitally slices the model into thin layers of approximately 0.005 inch (five thousandths), and produces a removable, stabilizing structure to support the part during the build. Next, the physical build process begins with a vat of photo-curable liquid resin and an elevator table in the vat, set just below the surface of the resin.

A computer-controlled optical scanning system directs the focused laser beam so that it solidifies the 2D cross section corresponding to the slice on the surface of the photo-curable liquid resin. The laser's depth of penetration is greater than the desired layer thickness, and is known as overcure. Overcure plays an important role in producing solid SL models and it also affects the build time of the part.

After a layer is complete, the elevator table lowers enough to cover the solid polymer with another layer of the liquid resin. A leveling wiper system moves across the surfaces to recoat the next layer of resin on the surface. The laser then traces the next layer. In simple terms, the energy of the laser "flips" the liquid material to a solid material upon contact. As chemistry buffs know, this is called a phase change in the resin. This process continues in successive layers, building the part from the ground up, until the system completes it. The elevator then rises from the vat, and the operator removes any excess, uncured liquid polymer from the part.

## STEREOLITHOGRAPHY (SLA)

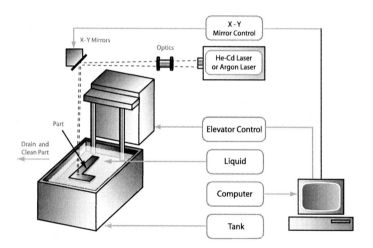

Finally, the part is placed in a UV oven for final curing. The part is then hand-finished to remove the support structure and to smooth the minute "stair-stepping effect" seen from building the part in multiple layers.

## STAIR STEPPING

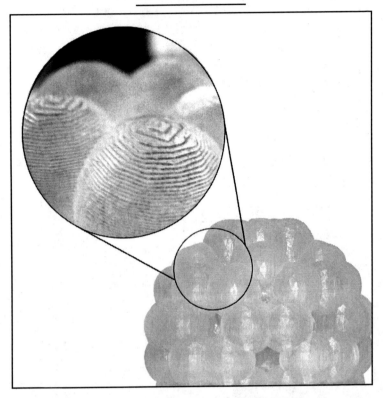

It is difficult to predict the cumulative impact of chemical properties and operating parameters on a build, subject to the ever-changing aspects of cross sections and geometries. Therefore, developing easy-to-use pricing algorithms has always been a challenge. The industry-accepted approach is to use geometric information in secondary formulae to predict cost and "guesstimate" the true build time of a part.

### ⁑ SL Applications—Stop and Smell the Plastic

Look around you. Almost everything you touch throughout your day was made using a specific prototyping process: cell phones, keyboards, pen caps, gearshift handles. Rapid prototyping has already reached every sector of your day-to-day life. As a result, a firefighter's mask fits better, your steering wheel grips tighter, and your office furniture looks smarter. The virtual-to-reality evolution makes room for organic shapes, compressed time cycles, and quick physical replicas. Home or office, work or play, most of the things you touch were designed and produced using the SL process.

## Industry Overview—A Nerd's Eye View

The innovative solutions made possible by SL technology are energizing the conventional design and engineering culture worldwide. According to the *Wohler's Report 2006,* in the last year, millions of parts were made by service providers and end-user companies combined. That's amazing growth for a teenage technology.

With SL's ability to produce finished solid objects within hours, the feedback loop is much faster, and approval time is shortened to days. In every boardroom around the globe, engineering managers are evangelizing the wonders of RP, as it frequently cuts tooling costs by 50% and reduces overall development times by as much. So the question is: **Why wouldn't you use RP?** The answer: lack of knowledge or corporate superstition. Maybe you tried it once and your expectations were not aligned with reality. In the world of RP, it's easy to ask for the wrong thing without knowing it.

In a young industry, product development leaders are very excited to see the best minds in design and engineering take ownership of these powerful tools and come up with a new generation of evolutionary applications. Pick up any manufacturing trade journal to find hundreds of fascinating case studies about SL applications across a wide spectrum of users.

## Resolution—The Nitty Gritty

Understanding part types as a function of resolution is very important as resolution affects tolerance, surface finish, and cost. As a customer, you have several choices of resolution when creating your SL part. Applications fall into three basic types of SL parts as defined by resolution: The standard layer thickness for SL parts is 0.005 to 0.006 inch (five to six thousandths), the thickness of a sheet of paper. High-resolution parts are grown at 0.002 to 0.004 inch (two to four thousandths), one-third as thick as a sheet of paper. Machines made by Objet Geometries produce the highest resolution layer thickness of 0.0006 inch (six ten-thousandths), not visible to the naked eye.

The resolution offered by three different process types dramatically impacts the accuracy and feature capability of your parts. Different geometries require a certain level of resolution. The thinner and more fragile part features are, the higher the resolution required. If your part has no highly detailed areas, standard resolution is sufficient.

If you are creating a part that is mostly simple with a few complex features, for example, a housing with buttons, you can build the complex features in high resolution and build the housing in standard resolution to realize a cost savings. While it would look great, building the entire assembly in high resolution would be considered overkill for the part type and would be very expensive. Typically speaking, the super-high resolution processes cost twice as much as the high-resolution process, and four times more than standard.

### Benefits—The Buzz is Real

Over the past five years, trade journals such as *Rapid Prototyping Report, The Edge,* and *Wohlers Report* have created a huge industry buzz by citing hundreds of success stories using SL. Whether you are designing a new engine block for Mercedes-Benz or a shoulder replacement for a human being, SL technology is radically compressing developmental cycles, saving millions of dollars, and opening new doors to innovative solutions. SL makes for almost magic stories with happily-ever-after endings.

One dental company heavily employs SL technology to create invisible braces to treat hundreds of thousands of patients. A major automotive company realized a cost savings of 45% by using SL for rapid tooling of small parts. A jet engine blade project, typically taking nine months with standard tooling and machining, took only one month using SL solutions. Design time for an orthopedic implant was reduced by 15 months. Design cycle times and production cycle times for an electric power system were reduced by 40%. One tooling project that would have cost $1,700,000 using conventional methods cost only $40,000 with SL because it drastically reduced the errors in tooling. Competing architects designing options for the new World Trade Center used SL extensively for highly detailed, full-color miniature models of their visions. A number of motion picture companies use 3D modeling and SL on a daily basis. To emphasize a sign of the times, the motion picture *Small Soldiers* featured an animated SL machine in its opening sequence, to the delight of product developers around the world.

SL technology continues to revolutionize manufacturing in all key industries of automotive, aerospace, military, machinery, biomedical and dental, consumer products, shoe-making, architectural, and aesthetic and artistic products. Additional industries along with entertainment and filmmaking include: forensics, space exploration, microsystems, geographical information systems, and mapping. The application of SL technology holds unlimited potential and is a challenge to every designer's imagination. It's easy to understand why **RP makes heroes.**

A day in the life of our humble hero, Johnny Quickparts, further details more advantages and limitations of the SL process. Let's stop by the perennially stressed-out Acme Design Corporation and see how Johnny is going to save his boss, Bob Overrun.

## A Very Tall Tale—Johnny Quickparts and the Olympic Torch Challenge

Deep in the basement of Acme Design Corporation, the product development team was playing solitaire and waiting for its next crisis. Fast asleep in his cubicle, the humble super-geek Johnny Quickparts, P.E., was having another work-related nightmare: to the drone of an antiquated air conditioner, Johnny watched as his engineering manager used medieval methods of hand-modeling to solve another emergency design problem. As the telephone rang, Johnny woke up, stretching and yawning, prepared for his next weird work experience. It was his boss Mr. Overrun in a panic.

The client needed a plastic model of an Olympic torch for a tradeshow by Monday or Acme Design would lose the account! If they drew it in 2D and whittled a prototype by hand, it would take three weeks to get approval. Mr. Overrun was willing, at last, to try Johnny's new "quantum thinking." Mr. Overrun pleaded, "If you can save this account, Johnny, I'll let you touch my Harley!"

Johnny rubbed the sleep from his eyes and began cogitating. He loved hanging up on his boss without saying a word. It made them both feel very manly. He opened a can of Pringles and thought of all the times he had seen the Olympic runner carry the early Grecian torch to start the games. Johnny swelled up with pride for finally connecting to the ancient past, the belly of Western Civilization where heroes were honored for their strength, flexibility, and speed, just like the little-known rapid prototyping processes he lauded. This was his one chance to perform an Olympic challenge, so he started a new 3D SolidWorks file, put on his headset, and cranked up the music, the theme from *Chariots of Fire*.

Atop the mist-covered Mount Olympus in his mind, Johnny drew an excellent 3D torch in about an hour. He wanted this torch to be way more than cones and cylinders. His torch would be curvy, freeform, and darn-near pretty. Johnny's torch would out-torch all others in wind-resistant and aerodynamic features! He had watched enough *Flintstones* episodes to make it organic-looking and ergonomically correct.

Toward completion, Johnny checked his file to make sure the data had no bad facets or holes. He hollowed it out to save on materials and build time. He made sure that there were no freaky undercuts or fragile features that might be difficult to build. Double-checking his work, he carefully eyed where the parting line for the eventual mold would be. Then he converted his CAD file to STL format.

Choking down a really bad cup of coffee, Johnny considered his options for making a physical torch, such as Computer Numerically Controlled (CNC) milling the part out of metal, foam, plastic, or wood, or using an additive process such as Selective Laser Sintering (SLS), Fused Deposition Modeling (FDM), or Stereolithography (SL). He chose SL over the other processes because he needed a smooth finish with metallic paint, something glossy. He was not really shopping for durable plastic. For a show model, he was willing to trade functionality for excellent appearance. He imagined that the SL process would deliver the goods quickly but he wasn't completely sure.

Suddenly the frosty Queen of Procurement, Sally Savealot, dressed in a tiger-print dress, graced Johnny's doorway. She eyed him up and down with her famous "engineering disdain" look, designed to vaporize the majority of problems and people that came her way. "For today's crisis," she spoke coolly, "just use a credit card and quit wasting my time!" Then she vanished.

Johnny knew she didn't mean it and submitted several requests for quotes (RFQs) the old-fashioned way. He then flew onto the World Wide Web and Googled Stereolithography, just for kicks. Then a truly mysterious thing happened. He noticed a company that shared his own last name...Quickparts! A believer in synchronicity, Johnny wondered if he was somehow related to a company that used all of his favorite technologies. Duty called and he could not tarry, so he shrugged off this name coincidence for the time being. He logged in and quickly uploaded his torch file to Quickparts.com and got an instant quote.

Within 10 seconds his phone jingled in an unusually warm, sweet way. It was Helen Helpalot from Quickparts, following up on his quote. When Johnny heard her velvet voice spiked with real enthusiasm, his left brain simply melted. She was already on his side and she would never lie or leave him. She even told Johnny that, as her customer, HE was indeed the heart of the part. Life was suddenly a blaring carnival for Johnny. He reached for his heavy gold crown to make sure it was on straight. Maybe he had inadvertently downed champagne instead of orange juice for breakfast.

In a matter of minutes, Helen guided him through his technical specs to make sure that SL was indeed the right process for him. Johnny shuddered at the ecstatic experience of dealing with this unusually wonderful sales rep. He loved the clarity of her lightning speed response. Not only was Johnny in love, but he would also get his torch the

## NEXT DAY! Bam! Done! Sold! This was big.

After the call, Johnny sank to his knees and took a long drag on a candy cigarette from his Halloween stash. He kept shaking his head, muttering, "Holy Cornflakes." This was a whole new universe of dynamic customer service with someone who really understood what he needed and actually cared about his part! By the time the other competing quotes finally came crawling in, Johnny was already on his way home to watch *Star Trek* reruns and think about Helen Helpalot. He almost dialed her up again but gave himself a noogie to snap out of it.

The next day at work, Johnny noticed the other quotes were bland and confusing. The prices were disparate and the lead times made no sense at all. None of the other service providers had even called him, and he wasn't going to call them. Clearly, Helen Helpalot was fueled up on the breakfast, lunch, and dinner of champions. He crunched the other quotes into a ball and slam-dunked them, just as the FedEx man brought in his SL package and grunted for Johnny's chicken-scratch. Johnny was so happy he threw his bony arms around the meaty, silent type in sunglasses who was apparently used to public displays of affection from total strangers.

Johnny now held a gleaming SL replica of a lifelike silver Olympic torch in his hands. He rolled it over and over in his palms, amazed at a vision made real. He could hear the roar of the coliseum crowd chanting,

### "Go Johnny go! SLA all the way!!!"

The smooth, metallic-looking torch looked and felt absolutely real.

With the torch held high above his head, Johnny sprinted 100 meters to Bob Overrun's mega-cubical. Mr. Overrun was beside himself with glee. The torch was way better than what he had envisioned. Johnny had saved his boss again and was now wearing the proverbial gold medal. He began to teach Mr. Overrun the basic magic behind his latest success, if for no other reason than to spice up his cocktail party repertoire.

Johnny was glad he had taken good notes from his conversation with Helen Helpalot. His shaky handwriting was a memento of their very first conversation but it would also help him justify his purchase to Sally Savealot in Procurement. After meeting Helen, Johnny considered abandoning his 10-year effort to win over Acme's Goddess of Purchase Orders. After all, his valentine of pink tulips had only made her tougher.

Johnny learned a great deal about the SL process from his torch project; and yes, the customer was triple-wowed. In fact, they ordered four more plastic torches for a fraction of the cost—thanks to "family build" savings! That night, Johnny went home satisfied and celebrated with a new tuna melt recipe for one. He kept thinking of the motto of the Olympics, *"Citius, Altius, Fortius,"* and how it relates to the world today.

## ▦ Boring but Necessary—Understanding Stereolithography Limitations

### *Part Features*

Knowing how to select an AF process for your part's feature limitations is very important. Every part is a series of features. Basic parts start off as simple prismatic shapes, such as cubes and cylinders. From these basic geometric shapes, a designer creates features. For example, a housing starts out as a box or cube and then is shelled out to reduce material. The CAD designer uses rounds, curves, and fillets to make the part more sculptural and organic.

Additional features are bosses, or attachment mechanisms, that interface with other parts. Features have thickness, fragility, and shape—characteristics that must be considered when using AF. So, if you are designing a diamond ring, you need a high-resolution process for thin, fragile prongs on the mounting piece. Those fine features steer you toward a high-resolution SL process that uses very thin layers of 0.002 inch (two thousandths) and steers you away from standard resolution manufacturing processes with a layer thickness of 0.005 inch (five thousandths). If you have robust features greater than 0.030 inch (thirty thousandths), a standard SL process is adequate. A wall thickness less than 0.030 inch is risky business because it won't fill up or form well in the SL process. A requirement for super-high resolution 0.0006 inch (six ten-thousandths), steers you toward the Objet process.

Unique part features are very important to consider when building a part. Typically, a part feature is integrated within the part and is not required to be analyzed independently for build or cost impact, but there are situations that require special consideration of the features. Some of these features include cylinders, such as tubes and cups, as well as rounded or sloped surfaces. How these features are handled will impact the cost of the part.

## Tolerances and Accuracy—Real World

The accuracy of the part is an important factor in building useful models. When using AF processes, there are typically no drawings or tolerance studies provided to determine whether a part is within tolerance. Therefore, the base dimensions are actual dimensions in the CAD model, which is mathematically perfect. The limiting constraint becomes the ability of the technology being used, such as SL, to produce the part.

In manufacturing, there is no such thing as a perfect part. Skillful engineers know it is necessary to apply tolerance, the permissible limit of variation in a dimension, to the design. What many engineers typically don't know is that SL is not an exact manufacturing process. Even the high-precision coordinate measurement machine (CMM) has tolerances. Its "1-inch" diameter pin is really 1.00005 inch.

With any manufacturing process, there are tolerances inherent to the process itself. Standard SL tolerances are ± 0.005 inch (five thousandths) for the first inch, and ± 0.002 inch (two thousandths), inch for inch on most parts and features. Understanding this is critical, especially when mating the parts made with the same SL process.

Tolerances and accuracy for SL are dependent on the geometry and orientation of the part. Engineers expect perfection because they design parts in the mathematically perfect world of CAD. However, once virtual comes into physical reality, we find that materials and geometries interact and affect the outcome. The perfect-fitting tolerance you had designed will most likely "behave badly" at a minute level in physical form due to residual stresses driven by part geometry. This is why **a designer must design to the process and material being used.** Minute residual stresses, determined by geometry, are introduced into an SL part when liquid turns into solid form. These residual stresses will cause the part to bend in a certain way. Outcomes can vary greatly from part to part, and process to process.

Of special consideration are those tolerances of two or more interfacing parts, produced in the same manufacturing process such as SL. Good engineers always specify the largest possible tolerance while maintaining proper functionality. But a clear understanding of how materials and geometries affect SL tolerances is most likely one thing they didn't teach in engineering school. Parts must be designed to distribute or relieve residual stresses. For example, a long bar will react differently than a housing part. While both designs look perfectly flat in CAD, they will both warp slightly, factoring together residual stresses, material properties, and build orientation.

A typical SL failure occurs due to the underestimation of tolerance stacking. Tolerance stacking occurs when mating more than one part because different parts have varying geometries that react differently rather than homogenously. While tolerances are somewhat less critical in the real world, applying special SL tolerances is highly critical, especially with mating SL parts where the interface may not mate at all due to distortion inherent in the process.

SL materials have a low tolerance for heat with typical heat deflection temperatures around 110 to 120°F. Tolerances may also be affected by splitting a part. Shrink factor is a component of accuracy and should be accounted for by your service provider operator.

### *Part Orientation—How to "Grok" XYZ*

Informed customers get better pricing and better solutions. To play this game well, you have to understand what orientation really means. The time required to build a part depends on its orientation in the machine vat while being produced. Factors include the number of layers required to be processed as well as specific layer-dependent parameters that can affect the time required to complete a layer.

Depending on part geometry, there can be a major cost and time difference in parts built vertically versus horizontally. Vertical builds get better definition and require a longer build time, and therefore, cost more. If you don't need perfection on your first draft, you may choose to build horizontally, and save 50%. However, when you are ready for a best quality SL, build in a vertical orientation to get better definition on your part.

> **QuickTip:** "GROK" (rhymes with "walk") is a fun, friendly science fiction verb, coined by author Robert Heinlein. It means "to understand something so well that it is fully absorbed into oneself." Example: If you can grok this important material, you will become a hero in your company!

There is also a tradeoff between the surface finish requirements of a part and its build time. Typically, surface finish is the more critical factor because a part with poor surface finish may not be useful to the user, regardless of how long it took to build the part.

### *Size Matters—Are We Surprised?*

In almost every category of life, size does indeed matter. Some of us have learned about machine size issues the hard way. 3D Systems machine names SLA-250 and SLA-500, actually denote their real build platform size. The SLA-250 platform is 250 x 250 millimeters, or approximately 10 x 10 inches. The SLA 500 platform measures 500 x 500 millimeters, which is approximately 20 x 20 inches. However, for the more recent models, SLA-5000 and SLA-7000, meaningful size denotations were dropped and a few zeros were added to make the machines sound "bigger and badder," though they are essentially the same as their predecessors.

The point is to be careful about machine size. So, if you have to produce a 21-inch rod in SL and you are dealing with a fly-by-night SL provider, he may suggest that, due to platform size limitation, he would cut the rod in half, build two SL parts, and then rejoin them with adhesive, a more costly solution. While it is true that SL machines offer only two choices of platform size, 10 x 10 inches or 20 x 20 inches, there is a better way to handle this problem. Look at the build platform space in a new way. Instead, turn the 21-inch rod diagonally in the larger vat so that it fits on the hypotenuse of the platform space. **Avoid splitting parts whenever possible.**

In other words, part size should determine machine selection. If your part has a dimension greater than 20 inches, you have to figure out how the part can best be oriented to make a single piece. Make sure that your service provider has a large-frame machine. If they only have a small machine, your part will be split, which adds cost. If you don't ask them about machine size, they probably will not tell you.

To produce parts as a single piece, the SLA 500 platform has the limitation of approximately 20 x 20 x 20 inches. But for larger parts, it is possible to split and join them after production with special adhesives and resin directly from the machine. For example, if you need to make a 40-inch tube in SL, cut your CAD model electronically and design the split with a special tongue-and-groove connection. An ordinary slacker would use only dowel pins as connectors, but a super-engineer takes the guesswork out of rejoining to make the tube halves fit perfectly together for durability and lasting quality. **Be proactive and determine the fate of your own parts.** Do not leave it up to a service provider who might just glue the halves together in a sloppy way.

A part is composed of many attributes, such as volume and the overall dimensions of the part. Obviously, these

> **QuickTip:** Make sure your service provider uses a large- frame SLA machine for big parts.

attributes have an impact on the part production cost, but there are extremes that have to be considered: very small parts and very large parts. Very small parts are, in essence, parts that require few resources to produce from the system. They require very small amounts of raw material, and their build times are very short in comparison to times for normal parts (measured in minutes). From a cost estimation evaluation, the cost of these parts will be insignificant and always default to some predetermined minimum part cost value, such as the direct cost of just starting the equipment.

The other extreme is very large parts, again very subjective. In SL, there are no constraints for the volume of the part. However, large volume parts could be outside the spectrum of the cost estimation algorithms. The value of the large volume would likely be in the range of 50 to 100 cubic inches, and all parts that exceed this limit would be subject to additional scrutiny in their cost estimation. Smart engineers know there is plenty of room to negotiate on big parts due to the huge variance in "guesstimates" on labor and materials. You can pay anywhere from $3,000 to $8,000 for the same part. Obviously, the primary factor affecting parts of this size is the actual build time to produce the parts, which can extend to days or even weeks. The best way to save money is to **negotiate with price matching from other lower quotes from equal quality providers.**

## Materials Are a Nightmare

Johnny Quickparts' boss, Bob Overrun, had a negative impression of SL. Ten years ago he had ordered an SL part, dropped it on the floor, and watched in slow motion as it shattered into a hundred pieces. Needless to say, his customer was not happy with the invisible part or the new industrial revolution. Johnny did his best to convince his boss that these once brittle materials have come a long way in a decade.

The truth about materials is that there is no truth. Materials are a nightmare. To push ahead of the competition, plastics companies continually release "new and improved" plastics and resins, making shopping very confusing. Plastics marketing professionals have a tough time pitching new products as stronger, stiffer, brighter, bouncier, or somehow sexier plastic. Basically, it's all gooey gunk. Some of it is toxic, some of it isn't. The performance of materials is geometrically dependent on design; orientation can determine the success of your build.

The best way to get to learn about materials is to feel samples with your own hands, compare their properties, talk to the experts, and try out a new material on your next project. The materials used by SLA equipment are mostly epoxy-based resins that offer strong, durable, and accurate models. These characteristics make SL an excellent all-around choice for prototypes. Informed users know that Objet Geometries typically uses an acrylic-based material that can be brittle and not that user-friendly. Some acrylic-based materials are potentially carcinogenic and less stable. However tempting, please do not eat the parts!

In materials selection, try to identify the material that supports the function of the prototype itself, which in turn can support the function of the part in the real world. Sometimes you will have to compromise, but companies offer a litany of SL materials to cover capabilities from rigid to durable to flexible. When shopping for plastics, you need to know the basic material types.

## Material Types—It's All Gunk

Rigid materials, similar to polystyrene or Acrylonitrile Butadiene Styrene (ABS)-like materials, are used for things like a computer mouse, cell phone, or electronic shroud. For harder parts that require no flexibility, rigid material is tough and can withstand rugged environments. In the industrial world, a handheld scanner that may be dropped or knocked around in an industrial environment needs a rigid material. Holding up to wear and tear, rigid resin ensures your part a long life.

Durable materials, closest to polypropylene, are used for parts that require a snap fit. Durable materials flex without breaking, but be cautious when building a part that requires flexing. Make sure that your part is oriented properly to support and strengthen the snap feature. A vertical build will add strength to a flexing snap feature, but the horizontal build is innately weaker due to horizontal layering. **Mistakes are commonly made due to lack of knowledge about part orientation.**

Semi-flexible material, like polyethylene, is lightweight and easily deforms. It is used for some bottles and lids. Flexible material, such as elastomeric, is rubbery and used for connecting pieces like gaskets, washers, and boots, which often require a watertight seal. Elastomeric is highly flexible and forms strong seals and interfaces. Water sealant can be added to make parts, such as flexible nozzles on liquid dispensers, water-resistant. Special materials are available for high-temperature SL usage.

Save yourself months of research by understanding that there are only a few basic materials that really exist. However, plastics and their distant relatives are marketed with more flavors and hype than Baskin-Robbins, Ben & Jerry's, and Häagen-Dazs combined.

## Water-Resistant—Glub, Glub

SL material will absorb liquid and cause it to deform. If you need a water-resistant SL, use water sealant as a secondary process. The best SL choice for water testing is the rigid material. Please note that the SL becomes water-resistant but not waterproof. Therefore, only limited use in water—an hour or two—is recommended.

## Temperature—Some Like it Hot, But Not Stereolithography

Standard SL materials are temperature sensitive and will not withstand more than 120 to 130°F before they start to breakdown, deform, and warp. In other words, don't leave SL parts in your car in the middle of summer. In only a few hours, those parts will twist and warp, as Johnny found out on a searing August afternoon. A quick errand turned into an extended sales pitch for Acme. His prize possession, a 1970 cherry red Camaro, got so hot that it torched his Velveeta sandwich and his SL part in about two hours. Thanks to extraordinary precognition, Johnny had ordered an extra SL part that cost only pennies thanks to economies of scale. No mistake was ever wasted on Johnny. He ate his drippy cheese sandwich and twisted the gooey SL gob into a perfect likeness of his dog, Attaboy. A resilient learner, Johnny would soon discover high-temperature materials used in SLS, allow parts to withstand temperatures up to 200°F.

## How Will My Part Look?

A critical part feature is the surface, meaning the part surface as it comes off the SL machine. You want it to look and function at its very best, so additional finishing, a physical alteration, is almost always required. Trained craftsmen do all post-processing hand-finishing of SL parts. Informed engineers know that finishing and post-processing involves taking the part off the SL machine, removing the support structures, and sanding down the part—all of which affect lead times. Both detailed parts and bigger parts take longer to sand, but typically finishing takes as long as building the part.

Prior to manufacturing, part orientation needs to be planned to eliminate the undesirable stair-stepping effect, evidence of the layering process. Therefore, it is important to orient the part to minimize the stair-stepping effect that all current AF systems produce. Stair-stepping is most often apparent on sloping or curved surfaces, but can also occur on flat surfaces, depending on part orientation. In certain cases, stair-stepping may be impossible to eliminate.

After manufacturing an SL part, sandpaper finishing is typically used to smooth a part's surface. Skilled craftsmen fill tiny holes and sand down cured SL material to get a smooth polish.

The importance of surface finish depends on the specific use of your part. Because craftsmen have different skills, the human factor is introduced here. An inexperienced operator might remove or "subtract" too much, whereas a skilled crafts-man will be more precise. **A craftsman's skill level can affect your part's tolerance and accuracy.**

Because there is a lot of science to a part, there are several causes of build failure. Bad CAD files and wrong orientation can crash a build. Machine parameters, such as truncated wait times between layers, can ruin the build. A power outage or a dead power supply also cause build failures. Johnny has personally observed that earthquakes, fires, floods, and temper tantrums never helped a build!

The cost of failure is absorbed by your service provider, but the loss of time hurts everyone. When a problem occurs, the whole part must be scrapped and rebuilt. That's why a service provider makes a resounding groan if a build that takes 10 hours fails in the last hour; the lost build time is non-recoverable all the way around.

### • Finishes

Most parts require a standard finish. Some finishes are more functional than aesthetic. The finishes listed here are used when you need an SL model to evaluate your part for some reason. As a customer, you have a choice of finishes.

Primed SL parts have several coats of automotive-grade primer applied. This coating makes the parts paint-ready or mold-ready. Primer is needed when you want to make sure your part is tradeshow quality. After being primed, you can paint the part yourself.

Painted SL parts have several layers of paint applied to color the parts to your specifications. This is needed for fully functionally show models that have been filled, primed, and painted to look like actual parts. Painted parts are used for aesthetic or illustrative purposes.

Painting your parts increases the service provider's price only because it takes a long time to paint them, not because of any special paint used on them. Paint is one of the best and easiest ways to color SL parts.

Make sure to ask your service provider if finishing supplies are automotive grade to provide the best finish possible. If you want to paint your own parts, be sure to use a sandable filler primer. Once the part is primed and sanded smooth, any type of paint will work fine. Parts must first be primed, prior to painting, so that the painted finish will look nice and last longer.

If you plan to submerge your SL parts in dyes to add color, keep in mind that SL parts do absorb liquid and can swell or warp under the conditions present in dyeing. Therefore, this method of coloring is not recommended.

The standard finish on SL parts is almost paint-ready. You could paint directly onto the standard surface; however, there are marks that will show through the paint unless you prepare the part by applying several coats of sandable filler primer.

Paint can be removed from an SL part. A quick bath in acetone or paint remover will begin the process; however, be sure to quickly wash the parts in water afterward to remove residual chemicals. Remember, SL parts absorb water. To completely remove all material, sand it away. Harsh chemicals will eat away at the SL material.

"Strip-and-ship" is not a psychological method for blasting obnoxious people out of your cubicle. Strip-and-ship refers to SL parts that look ugly because they have no finishing, other than the removal of support structures. Informed customers who are price sensitive can typically negotiate a 20% discount for this finishing level. If you are a cheapskate, like Johnny's boss Bob Overrun, the kind who doesn't want to pay for anything, please remember that you are asking for an unfinished part taken right off the machine. Strip-and-ship is not recommended because customers typically under-value all the labor that goes into making a standard finish. **With strip-and-ship, beware: what you don't see is what you get.**

The last choice of finish is WaterClear, a high polishing process. WaterClear gives white opaque plastic a mirror-smooth, clear-looking finish, needed for glass-like parts such as LEDs, bottles, lenses, and covers. WaterClear material cannot be tinted but it can be made optically clear, although it may have a slight yellowish hue.

### How Long Does it Take to Get My Part?

Johnny hated it when Mr. Overrun yawned repeatedly during zealous reports of technical discoveries. Overrun would always interrupt and slur a question through the dark cavern of his yawning mouth, "Okay, Johnny, so how long does it take to get my parts?" We all want to know this: how quick is quick? Is rapid really rapid?

Typical lead times for SL parts are as follows. Most standard SL part orders are delivered in 3 to 10 days, depending on the size of your order and your service provider. WaterClear finishing and painted parts take a bit longer, approximately 7 to 14 days.

**QuickTip:** In every case, companies save significant money by using their own shipping account information. Service providers buy bulk rates for shipping but charge you the commercial rate.

High-resolution parts also take a bit longer, approximately 5 to 14 days. Some service providers do offer next day shipping if you need parts fast.

If you are not in a rush, you may negotiate a 25% cost savings off the standard shipping cost if you order economy shipping service.

Along with cutting lead times and reducing your time to market, plan well in advance and extend lead times to match your real needs. Allowing your service provider longer lead times can save you money.

## ⠿ Saving Money—Saving Time—Saving Thousands with SL

> *"Computers aren't emotional. That's why*
> *instant quoting is good. They won't raise*
> *your price because of a bad mood."*

After reading this book, you won't be a technical expert, but you will be an expert in saving time and money. No other book tells you how to be a hero using these processes. Here are some insider secrets to saving time and money using SL.

### *How Do I Save Money Using SL?*

Fortunately, SL has some very powerful characteristics to help drive efficiencies which can save money.

### • Family Build Concept

A powerful cost-saving secret of using SL is realized using economies of scale. Economies of scale refer to the decreased per-unit cost as output increases. In other words, the initial investment of capital is spread over an increasing number of units of output, and therefore, the marginal cost of producing a part decreases as production increases.

Engineering managers can save thousands of dollars using family builds. Be prepared to use this powerful knowledge when negotiating with service providers that may offer only traditional pricing. Teach your provider about economies of scale in the following way.

### • How Family Build Works

A service provider's operational overhead is built into every minute of the manufacturing process. Machines do things step by step. In between those steps, there is wait time. For example, in SLA production, the vat of resin takes 30 to 60 seconds to settle after the elevator platform moves to create the next layer of your physical model. This wait time is a valuable resource that is either captured or becomes operational waste.

If you produce a single part in that vat, the wait time or operational waste is the same as if you run a family build of 10 different parts in the vat. But in the case of the family build scenario, that wait time is divided by the number of parts in the vat. Therefore, operational waste is much lower per part than when running a single part. Traditional pricing models would show that if a single part costs $200 then 10 parts x $200 each would cost $2,000.

With economies of scale, pricing is much less per unit. Imagine your single part costs $200 to produce. If you build the family of 10 different parts, the cost is $425 total! Using family build, your piece part price decreases from $200 to $42.50 because operational overhead is now distributed among all the parts.

**QuickTip:** Buy a few more, save a lot more. Learn the magic of family build!

A service provider using economy of scale pricing is very advantageous to those who need lots of different parts. Track your quotes from service providers to see if the single part is much higher than pricing for groups of parts. This knowledge will also help you understand the often puzzling disparity between quotes from different providers. Look closely: A cheaper price per unit may be resulting from economies of scale, but our traditional thinking would equate a lower price with lower quality. However, this is not the case when using family build. A lower price will most likely get you quality equal to or better than other competing quotes.

## • Build Multiple Parts of Assembly Together

A good way to capture family build savings is to identify multiple projects needing assemblies of multiple parts, such as cell phones, keyboards, or a housing with buttons. If all of the parts fit on the SLA platform, you can save up to 50% of your traditional pricing estimate.

## • Order Multiple Quantities

Economy of scale savings also apply to ordering more than one of the same part. Imagine that you need a pen cap made in SL, costing $200. But others in the company, like marketing and the president, also need one. If you order more than one, each additional part will be dramatically lower than the first. Traditional pricing would say that three pen caps x $200 each is $600 total. A better family build price shows the first pen cap at $200, but three parts together are quoted as $250 total, or a piece part price of $75 each.

## • Use Single Material

Try to use a single material when building SL parts to save money. Family build practices can extract all overhead from one material. If additional materials are used, overhead will be added to those as well. So, the overhead is the same for either one or more than one part when using the same material during the same build.

## • Orient-to-Fit

Know the difference between building parts vertically versus horizontally. Vertical builds take longer, have more definition (based on geometry), and cost more. Be creative in fitting your part to the build platform for best result and cost. **Knowledge of part orientation will help you get the best deal.**

## • Instant Quoting

Find sources that use instant quoting to literally save your company hundreds of man-hours of labor. Always use instant quoting for pricing because it has no greed factor and no human element. Instant quoting actually saves you money before you buy by helping you manage your quotes. An engineer's time costs a company about $100 per hour, so if you spend six hours getting a quote, you are losing money. If you decide to wait it out and get old-fashioned manual quotes, be sure to study them for mysterious markups.

## • Other Ways to Save

Responding to the new age of quantum manufacturing, many companies hire an RP coordinator to assess departmental needs and place consolidated orders, saving their company thousands of dollars. They maximize savings with service providers by combining multiple projects when ordering SL parts, buying fixed hours on machine time in bulk, getting preference for volume discount by buying everything from one service provider.

## *How Do I Waste Money in the SL Environment?*

It's easy to inadvertently waste money in the SL process if you are a newbie. Here are ways to avoid wasting money.

## • Wrong Orientation

By now you know that wrong part orientation can foil your best intentions. Remember that vertical builds get you more definition and take longer, but horizontal builds cost quite a bit less and are good enough quality for a first draft. Keep in mind that building in the wrong orientation will result in an unusable part.

### • Build the Whole Thing

Regardless of the part size, the natural tendency of engineers is to build the whole assembly when all they really need is the functional part. Only build the pieces you really need. Be conscientious of what feature set you are trying to identify with the part. For example, with a computer monitor frame, you may only need to test the button area. The solution is to electronically slice your CAD model to produce an SL of the featured button panel.

### • Ordering "Onesies"

Single part ordering is not efficient. Operational waste racks up the cost very quickly, so plan ahead for multiple part orders.

### • Use Multiple Materials

Switching materials during your process is very costly, due to operational waste incurred. Keep it simple and use one material at a time.

## *How Do I Save Time Using SL?*

There are many easy things the engineer can do to save time, such as understand tolerances and materials.

### • Tolerances

In the CAD world all things are perfect; in the SL world, a part can be off 0.005 inch (five thousandths) for every inch of the part. Tolerances affect interfacing parts. In your CAD design your two parts fit perfectly together, but once manufactured they won't fit perfectly. The engineer blames the SLA machine, but it was their responsibility to adjust for SL tolerances. Knowing this can save you puzzling mistakes and timely rework.

### • Know Your Materials

For the SL process, investigate material choices thoroughly before you commit to one and request samples and data sheets well in advance of your need.

### • Double-Check CAD

Make sure that your CAD file has smooth surfaces before uploading it for your service provider. Faceted files can crash your build and waste time and money. Also, as commonsensical as it may sound, always make sure you send the correct revision of your CAD file. Many times when engineers are burning the midnight oil, they send the previous revision to the service provider, and get a nasty surprise when the part arrives.

### • Misnomers

Many people call all AF processes SL or SLA, even though they are referring to other processes such as FDM or SLS. Because SL has become an industry slang term for any AF technology, be sure you know the difference and call a process by its correct name.

### • Use Instant Quoting

This incredible software tool actually saves you money before you buy. As mentioned earlier, the quoting process can eat up serious engineering man-hours. Imagine saving four to eight hours every time you need to quote on something. That's saving hundreds of hours and thousands of dollars per year in the quoting process alone!

## *How Do I Waste Time in the SL Environment?*

The biggest time eaters in getting your SL made are caused by a lack of knowledge, which almost always results in costly rework. The safest approach is to increase your knowledge and avoid rushing through the steps for uploading files. To review, the most common time wasters are selecting the wrong material for the process, sending bad CAD data or STL files, building the wrong part version, based on a previous revision of a CAD file, using old-fashioned manual quoting, and selecting the wrong process for part.

## ⠿ The Keys to a Brand New Ferrari

By now you are revved up about SL technology and the dramatic time and cost savings it offers. You've heard numerous case studies of 50% savings or more with SL. You've made friends with our humble hero, Johnny Quickparts, the geek without guile who makes a dazzling technical contribution to his ho-hum corporation. You've seen Johnny beat a trade show deadline and use SL to verify his concept early in the design game, thereby eliminating expensive design changes late in the manufacturing process. You know that orientation, resolution, and tolerance are essential to building your SL parts successfully. You also know insider secrets on industry pricing. Finally, you've been forewarned about all those time eaters and shoulda-coulda-wouldas.

As Johnny would say,

**"Let's drive this thing!"**

# Chapter 3

# Selective Laser Sintering

## POWDER TO THE PEOPLE!

*"SLS is evolutionary low-volume manufacturing."*

### *QuickSMART*

**D**efinition: Selective Laser Sintering (SLS) is an AF process based on free-form technology and is used to create prototypes and functional parts. This AF method creates solid 3D objects by fusing or sintering particles of powdered material with a hot $CO_2$ laser. A thin layer of powdered thermoplastic material is rolled onto a heated build platform. The laser beam directs cross-sectioned CAD data onto the surface of the powder bed. The heat laser then traces each layer, melting plastic particles to the previous plastic layer. After each cross section is scanned, the powder bed is lowered by one layer thickness, then a new layer of material is applied on top. The process is repeated until the part is complete.

**Why You Need It**: To skip the prototyping process and directly manufacture end-use, functional parts; to get durable, heat-resistant parts (200 to 300°F); to quickly make parts for use in tough environments; to test parts in a real environment; and to create complex geometries.

**Ideal Uses**: Durable, thick, bulky parts, such as engine blocks, engine components, mounting brackets, and hot liquid dispenser components; trade show models, and concept models for reviewing ideas, form, and style; functional models and working prototypes; master patterns, investment and sand-casting patterns; heavy industrial use in automotive and aerospace; durable low-volume production parts; snap fits and living hinges; hot airflow model, used for testing.

## SLS Background

Borderline magic, SLS turns powder into parts in a matter of hours, typically building at a rate of one cubic inch per hour. SLS technology is widely used around the world for its ability to produce complex, durable, functional parts directly from a digital CAD model. While SLS began as a way to build prototype parts early in the design cycle, it is now being used in low-volume manufacturing to produce strong, fully functional parts with an accuracy of 0.005 inch (five thousandths).

Originated by DTM Corporation, SLS was acquired by 3D Systems, Inc., in 2001. 3D Systems, based in Rockhill, South Carolina, now manufactures and sells SLS systems and materials worldwide. At present, there are only a few known powder-based AF systems and 3D printers.

Another major player in SLS equipment is EOS (Electro Optical System) GmbH of Munich, Germany. While technically better, EOS leads SLS sales only in Europe. Leading sales in the US, 3D Systems has benefited from replicating many features of the EOS system. EOS continues to manufacture laser sintering systems that are quickly developing the future with variants of materials combined with polymers and metal for a wide range of production and foundry applications.

## SLS Process—Inside the Magic Oven

SLS offers the key advantage of making functional parts in end-use materials, such as ABS-like plastic. However, the system is mechanically more complex than the SLA and most other AF technologies. A variety of thermoplastic materials such as nylon, glass-filled nylon, and polystyrene are used in the SLS process. Surface finishes and accuracy are not quite as polished as those made with SL, but material properties can be quite close to those of the end-use materials. The SLS method has also been extended to provide direct fabrication of metal and ceramic objects and tools. SLS uses a hot $CO_2$ laser to sinter powder-based materials together, layer-by-layer, to form a solid 3D object. Equipment for the SLS system consists of a sealed chamber containing the build envelope, roller, and pistons; a $CO_2$ laser; and a scanning system.

### *Step by Step with SLS*

First, an operator converts your CAD file to a Standard Tessellation Language (STL) file, then the SLS system software processes the file and orients the part for optimum build. Next, the SLS software slices the STL file into electronic layers and sends it as instructions to direct the operation.

Thermoplastic powder is spread by a roller over the surface of a build cylinder. The piston in the cylinder moves down one layer thickness to accommodate each new layer of powder. The powder delivery system is similar in function to the build cylinder. Here, a piston moves upward incrementally to supply a measured quantity of powder for each layer.

A laser beam is then traced over the surface of this compacted powder to selectively melt and bond it to form a thin layer of the object. The fabrication chamber is maintained at a temperature just below the melting point of the powder so that heat from the laser raises the temperature slightly to cause sintering. A nitrogen atmosphere inside the fabrication chamber prevents the material from burning. The process is repeated until the entire object is fabricated.

## SELECTIVE LASER SINTERING (SLS)

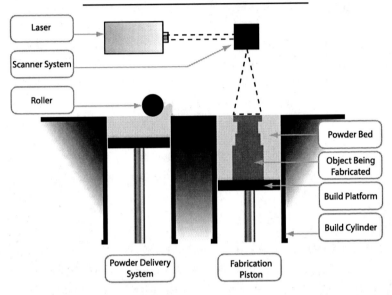

The remaining unsintered powder surrounds the object and acts as support for the part. Therefore, **no additional supports are required with SLS** since more fragile features, such as overhangs and undercuts, are supported by the powder bed. The SLS machine must cool down before the part can be removed from the machine. Large parts with thin sections may require as much as two days of cooling time.

After the object is completely built and the machine has cooled down, the part chamber is moved to a breakout station. Technicians excavate the part from a mound of powder. Excess powder is brushed away and manual sanding smoothes the unavoidable stair-stepping effect.

Sintered objects are porous. Therefore, it may be necessary to infiltrate the part, especially metal composites, with another material to improve mechanical characteristics. SLS parts are typically dipped in a "super glue" to fill the voids and make the parts smoother and more durable.

## ▓ SLS Applications—Shift Happens

Look around you. A profound, paradigm shift is happening, almost beyond our awareness. Some people call this a "sea change," a great metaphor for this sudden technology shift. Have you ever been sailing on the open sea when weather comes up rapidly? The water turns from a clear aquamarine to impenetrable darkness in what feels like seconds. The wind comes out of hiding, as if it had always been blustery. The change was gradual but you weren't watching every micro-movement of clouds that changed your world. In the same way, we tend to notice the accumulation of change all at once. Suddenly everything looks and feels entirely different. A new reality leaps into our awareness and we adapt to it as quickly as possible. This is how technology moves. Suddenly the coffee shops are filled with customers talking on their wireless headsets—as if things had always been that way. Nothing can undo these shifts except the next shift.

Layered manufacturing for low volumes has made it possible to experience a manufacturing "sea change" in our generation. Seemingly, all of a sudden, low-volume production of 100 to 30,000 parts is available as a reasonable manufacturing strategy for product development. With the enhancements in the technologies and materials, the use of layered manufacturing makes it possible to skip rapid prototyping (RP) and go directly to end-use, functional parts. Maybe our culture of instant gratification has driven us to innovation. You want it now. You get it now.

### Industry Overview—A Nerd's Eye View of SLS

SLS is a dynamically growing industry because the world wants mass customization now. Designers want parts faster. Managers want to save money by utilizing a process that allows them to make parts without the tooling time and expense. SLS means that you can shave weeks off the traditional product development cycle so that the process yields greater flexibility in the products you produce. The technology of SLS provides a basis for producing functional parts that withstand heat and environmental conditions, making it the clear choice for high-heat and chemically-resistant applications.

SLS has a strong reputation for making parts that can be used in a real environment. By using strong plastic materials—nylon—that provide very good mechanical properties, these parts bridge the chasm from being mere prototypes to being functional parts. An intake manifold is an excellent application for SLS. The heat-resistant SLS part can effectively be airflow-tested in an engine's exhaust system. The SLS part will be durable enough to be mounted on the engine and used in its real environment. Door handles are another good application for SLS because that SLS withstands a lot of force. Despite repeated use in the real world, these tough parts won't break.

While laser sintering technology and materials have not changed much in 15 years, EOS GmbH is making some exciting headway with experiments in composite materials. To really move the technology to the next level, materials have to become much more functional and practical. Parts have to resemble real-life parts. Unfortunately, a real part "look and feel" is still missing from the process. The technology and materials used are close to delivering real parts but are not quite there yet.

Engineers can view SLS as an option that allows for more creative thinking. Because they can get functional parts directly from CAD, engineers can cut guesswork out with revisions early in the design cycle. SLS is also an excellent solution for low-volume production of parts, meaning 100 to 30,000 parts. It is not recommended for high volume because, at a certain level, it makes more sense cost-wise to build a tool and produce from that to lower your price per part.

To understand the AF market competition, think of the competing sodas: Coke, Pepsi, and a third place cola. Like Coke,

SL leads the parts market, with FDM and SLS in a close race for the second and third slot. A deeper knowledge of each process will help you choose the best way to support your thirst for innovative design.

## Resolution—The Nitty Gritty

Because the SLS laser beam spot size is relatively large, standard resolution for SLS is typically ± 0.005 to 0.006 inch (five to six thousandths) for the first inch, and ± 0.003 inch (three thousandths) for each additional inch, with a layer thickness of 0.004 inch (four thousandths).

With the additive process, the z height (vertical axis) standard tolerances of ± 0.01 inch (one ten-thousandth) will impact the first inch, and ± 0.003 inch (three thousandths) for every inch thereafter. Orientation defines how these will impact the part.

Unlike SL, with SLS there is no option for higher or super-high resolution. The spot size of the laser beam reduces the precision of the SLS system which cannot produce fine features. Being a heat-based system, the SLS material expands and contracts with changes in temperature, causing warpage and tolerance issues at a microscopic level.

## Benefits—The Buzz is Real

SLS is the bridging technology to get your parts right now without tooling development and cost. It allows you to skip the prototyping process and get your hands on durable, real parts fast. The chief benefits of SLS are tougher, heat-resistant, functional parts that embrace complex geometry. Thermoplastics used in SLS are also easily bondable and machinable, being less brittle than SL materials.

While SL encourages optimization of a 2D build platform, SLS features a full 3D build envelope. Parts can be "nested" electronically, so that boxes can be built inside boxes inside boxes. A fully utilized 3D build cube and its ability to utilize powdered materials make for fast production throughput. Since SLS does not require support structures, post-processing is already at a minimum.

A key advantage of SLS over SL revolves around material properties. Evolving into greater flexibility with new composites, SLS companies are now experimenting with a wide range of materials that approximate the properties of thermoplastics, composites, nylon, glass-filled nylon, metal composites, and ceramics.

To get the best result with SLS, remember that it favors thick, prismatic parts and does well with complex geometries if they are not too thin. Fine features, thin walls, and organic details are a no-no with this somewhat brutish process designed to make thick parts that last. SLS does not need post-curing but it does have a long cool-down period. A design with thin walls can be problematic, resulting in distortion and longer cooling times.

While the technology has not changed that much since its inception, SLS makes tough, durable parts that can be used in production. These parts and prototypes can withstand high temperatures; therefore, SLS is used frequently in aerospace and automotive applications where, for example, an engine part may be mounted and tested in its real environment. An SLS part is also very useful if you need to test a design for a hot liquid dispenser because it can stand the heat. Well-documented in manufacturing trade journals, here are some brief examples of SLS applications from industry leaders.

A major automotive company used SLS to compress cylinder head development time from 16 weeks to 4 weeks and reduce cost from $75,000 down to $12,000. A military jet project containing over 80 SLS parts makes full use of rapid technologies. Race car manufacturers, including Formula 1 teams, have used SLS for several years to make diverse parts such as housings and aerodynamic components. A high-end luxury car maker eliminated the need for an expensive injection molding tool by using SLS to produce parts for a window lifter assembly. Many case studies report the use of SLS for medical purposes such as hearing aids or prosthetics. Consumer product and electronics design is the largest industry sector using SLS, making up one-quarter of the SLS applications pie. Medical, automotive, and aerospace applications trail close behind as top industry users.

The future is bright for SLS. Because of part durability, it's a natural foundation of layered manufacturing and the future of custom plastic manufacturing. Reminiscent of the cartoon world of *The Jetsons*, this new technology will soon be able to reproduce replacement parts directly from the machine that needs one. In fact, space colonization how-to is based on using machines that can "build themselves" once they are shipped to the moon.

**SLS applications are unlimited.** Wander through the superstores and study the kitchen gadgets. It's all plastic, it's all relevant, and it's all good. You never know what invention this field trip may inspire.

As we head into mass customization in the product market, the flexibility of SLS will continue to reduce time and cost in product development and eventually allow the consumer more variety at or near the point of purchase.

Let's stop by Acme Design Corporation to see how our gentle hero Johnny Quickparts uses SLS to promote world peace, but only after his power nap.

### A Very Tall Tale—Johnny Quickparts and the Night of 1,001 Brackets

This time, Mr. Overrun was really scared. Something very strange had come up. A new clip design had to be submitted the next morning or, as Mr. Overrun believed based on a turbulent voice mail, his own head would be FedExed to the Sahara Desert.

The customer, a tyrannical computer manufacturer, in the tiny village of Mama-Khan, shipped computers by rail, which shook the daylights out of them. To make matters worse, the train would often get stopped by a sandstorm and sit baking under a fierce sun. Due to a failed clip, the printed circuit board internals were jumbled and melted by the time they arrived at the end-customer in the desolate village of Daddi-Khan-Tu. Then the entire emotional transaction would start all over again. Needless to say, billing disputes were mounting and bad blood was flowing back and forth between the small but passionate villages.

The clip designed to hold computer internals in place had failed nine train rides out of ten, but both parties continued to hope. This SSDD (same stuff, different day) rationale led to a modern-day feud between Mama-Khan and Daddi-Khan-Tu, involving indignant extended families and carrier pigeons that dropped mildly insulting notes back and forth.

Mr. Overrun begged on bended knee, "This is my absolute last emergency—I swear. It is a matter of life and death." Overrun had never pulled the Death Card before, so Johnny replied, "Okay, but how about a large Everything Pizza and a Big Gulp?" Mr. Overrun took off like a laser-guided missile.

Johnny loved the notion of saving Mr. Overrun's head, promoting world peace, and solving another impossible technical challenge, all in one whack. He donned his magic creation cap and left a voice mail for Sally Savealot in Procurement. He was surprised to hear himself barking like a tough manager to grease his purchase order. She just didn't get it that he was a technological Elvis.

Johnny munched a stack of Pringles while his engineering mind soared across an expanse of golden sand dunes. He put on his head phones, cranked up the theme to *Lawrence of Arabia* and began a new CAD file. "Freaky-deaky,"

Johnny exclaimed. This was indeed an impossible order. The customer needed a design in the morning and 1,001 brackets in a week. Acme typically needed six weeks and $10,000 to make a tool, plus $5 per plastic part. Johnny had to find a way to make 1,001 parts without going through the tooling process. His new clip also had to be heat-loving, train-friendly, and durable. This was serious pressure.

Johnny imagined himself swathed in flowing white cotton, riding a stately camel around the trainload of jiggled computers. He could imagine exactly what the customer needed: a durable, heat-resistant plastic bracket to hold down the printed circuit board (PCB) inside the computer cases. It would have a very cool shape as well.

Johnny was thrilled to have a bona fide reason to call Helen Helpalot of Quickparts. In her special way, Helen confirmed his superior choice for this unique application: SLS was an excellent choice for mounting PCBs in a super-hot computer chassis, using nylon to low-volume manufacture without tooling. Then she added, "Johnny, you really understand that layered manufacturing for low-volume production is more than just parts; it's a way of *seeing*. Did anyone ever tell you that you are really special?" Johnny choked and managed to squeak out a few words, "Only my Grandma but she has distorted non-rational thinking." They laughed their first belly laugh together which made the shy bachelor really nervous. He scraped himself off the ceiling and got back on his camel. It was time to work. The clamoring harem would just have to wait.

After the call, Johnny took a long drag on his candy cigarette. That Helen could fill up his senses like Christmas, Easter, and Halloween, all rolled into one. Just guessing her hair color unraveled him into a state of grinning non-productivity.

Johnny turned the failed PCB clip over and over in his hand until he became the part. All of a sudden he felt weak and unstable, being thin, flat, and poorly designed in two mating halves. If he squinted his eyes, he could see into the misty past; the designer of the old part was so low on endorphins that he had used two halves to accommodate the constraints of tooling—not a visionary way to design. In the old world, producing two pieces would have doubled his tooling cost.

Johnny called upon his creative powers to find a design combining grace and efficiency. His new clip design would rock from concept to materials. Those people in Daddi-Khan-Tu would smile ear to ear when they inspected the snug-fitting goods. Being a true hero, Johnny wanted the two villages to live in peace at last.

Knowing the material limitations of SLS, Johnny could only imagine the old clip design bowing like a Tupperware lid in the fierce Sahara sun. He decided to design a complex U-shaped clip, merging two parts into one. This would not only reduce distortion but add strength. He knew that SLS could handle it, since it was friendly to complex geometries.

By the time Johnny had finished designing, he realized he had been singing, "I Did It My Way" all night long. The first rooster crowed in a nearby cornfield. He emailed his new clip design file to Mama-Khan for the fly-through. He slurped down his last bag of Tang for Astronauts and waited for a happy email approving the CAD model for low-volume production parts in seven days.

Johnny recounted the beauty of his latest job's impact. If he were the customer, he'd be ecstatic to discover that the hefty tooling cost had been eliminated completely. He would also consider it pure magic that he could get his 1,001 parts in only seven days. Thanks to layered manufacturing, the customer would pay only $25 per SLS part and save six weeks on tooling. This would generate a million dollars in revenue by getting it to market that much quicker. He could see that engineers really do affect business.

Just then, a new email flew in, commanding Genius Johnny to

### "Go forth and multiply my parts!"

With the click of the mouse, Johnny gave the green light to Helen Helpalot at Quickparts then stumbled home to his most exotic sleep ever. Johnny Quickparts dreamed of silk tents in the desert, a celebration feast of pomegranates and barbequed goat, happy sheiks with laptops, and super-approachable belly-dancers, all in a day's work.

---

## :: Boring but Necessary—Understanding SLS Limitations

In all AF processes, resolution is very important because all of the parts are created from minute layers. The resolution or detail that can be attained for each layer accumulates from layer to layer, and drives the output of the final part.

Because the SLS laser beam is relatively large, standard resolution for SLS is typically 0.005 to 0.006 inch (five to six thousandths). There is no option for higher or super-high resolution. As mentioned previously, these durable SLS parts can withstand high temperatures. However, part geometry determines the result since it is a heat-based process. Geometry will determine whether a part curls up or keeps its form. For example, a large flat part such as a cookie tray would bow up due to the residual stresses of heat in the process, whereas thick parts would typically do well. **Avoid using SLS for super-detailed parts or large, flat parts.** Mountable flat parts with bolt holes can work using this process. During the build, parts will warp slightly, but once mounted and fastened, the part will flatten and stay in place from pressure.

## Part Features

Features smaller than 0.005 inch (five thousandths) will not build well using SLS. If your part is finely detailed, you may want to consider other options. Features less than 0.030 inch (thirty thousandths) are considered high-risk. Think of SLS as a "man's man" process, making **tough parts for tough environments.**

## Tolerances and Accuracy—Real World

SLS is not nearly as accurate as SL. Accuracy for SLS ranges between 0.010 to 0.020 inch (ten to twenty thousandths). Parts that mate will need tolerances to support this difference in SLS. As previously mentioned, heat makes it difficult to control material properties. Don't expect tight tolerances in SLS, and be sure to adjust your design accordingly.

## Part Orientation—How to "Grok" XYZ

As with SL, part orientation is very important. Orientation in SLS also determines the definition of the layers and affects the resulting features. For example, a spherical part with a circular cross section must be oriented perpendicular to the laser beam to keep its roundness. Because SLS does not have support structures like SL, there is no orientation concern for supports.

## Size Matters—Are We Surprised?

Size options vary with SLS based on vat size and shape. Parts must fit onto build platforms measuring 11 x 13 x 17 inches. If you want to build an uncut, single-piece part, its dimensions must be less than 10 square inches. Otherwise, it will have to be electronically split, built in two parts, and then rejoined after the build. If your parts are large, the typical option is to split and rejoin; however, there are options. Service providers should be able to advise you on how splits are geometry-driven.

For serious designers and engineers ready to get started, please note that while the maximum dimension for instant quoting is 11 x 13 x 17 inches, the SLS build envelope is much bigger than that. For bigger parts, you may consider SL as an option. There will be some manual guesstimating in the quoting process related to splitting and rejoining pieces. When splitting a part, remember that thin parts are difficult; blocky parts work well.

Because SLS features a fully 3D build envelope, opportunities for orientation and creative nesting of parts abound. Here's a simple way to explain nesting to your grandma: just imagine a magic oven that can stack and bake 20 cookie sheets all at once. Don't confuse her by telling her that she could bake small cookies inside of bigger cookies, all at different angles.

### Material Types—It's All Powder

The good news is that the SLS process provides one of the most functional AF parts available. The bad news is that these powdered materials and their variations are about as exciting as competing aspirin labels.

Material choices for SLS include Duraform (a polyamide nylon material), DuraformGF (glass-filled), Somos 201 (flexible, rubberlike), and Castform (wax). If that means absolutely nothing to you, don't panic. All you really need to remember is that SLS uses a variety of plastic materials and binders to produce parts. Parts made by SLS are about 10 times tougher than SL parts and the same strength as FDM parts. Because of porosity at a microscopic level, SLS is not water-resistant. A super glue-like sealant is needed in finishing.

### How Will My Part Look?

Don't be mad or sad if you open your FedEx box only to find that your SLS parts are not very pretty. SLS parts have a sandy, porous surface. While SL can be processed and painted to look aesthetically pleasing, SLS does not take kindly to sanding, priming, or painting. It *could* be done if it had to be done, in the same way that you *could* fit a refrigerator in a car trunk, but only if you are really looking for that kind of fun.

Because there are no support structures with SLS, no strip-and-ship option is available. These parts are always sanded before delivery into customer hands. Because SLS is naturally rough, it can be sanded smooth but not sleek. Although it affects tolerance, SLS parts are dipped into a "super glue" to make them stronger and smoother.

**• Finishes**

With SL there are a number of options for finishing, but with SLS you get a standard, sanded finish. SLS is the functional workhorse, while SL is the show pony.

### How Long Does it Take to Get My Part?

Like SL parts, delivery of most SLS parts is fast—typically three to five working days. But, depending on the part size and complexity, your parts can take weeks to build. There is no Next Day offered with SLS. While parts do build fast, the machine takes hours to cool before the parts can be removed. With SL, there is no cool-down time.

With SLS, you can have build failures, but you won't know it until the part is completely built then excavated out of a pile of white plastic powder. Imagine an archaeologist carefully brushing away small amounts of dust until the treasure, your part, is revealed. With SLS post-processing, take extra care to avoid damaging the part. With the SL process, you can watch as the part is building. With SLS, you are going on blind faith until it is ready, hoping that the part turns out well. But take heart—the failure rate of SLS is low, so it is not a huge concern.

### ⠿ Saving Time, Saving Money—Saving Thousands with SLS

### How Do I Save Money Using SLS?

Although SLS costs approximately 20% more than SL, its advantages more than justify the cost increase. The SLS process typically saves the entire cost of tooling; therefore, the cost per part is justifiably higher.

## • Family Build Concept

As mentioned in the SL discussion, economies of scale also relate to SLS. While SL builds on a 2D platform, SLS builds in a fully 3D envelope so parts can be run with high efficiency. The operational waste is distributed among all the parts in the vat: the more parts you build at one time, the less they cost. A good way to visualize the 3D envelope is to remember the magic oven that stacks 20 cookie sheets fully loaded, all at the same time.

## • Multiple Quantity Part Orders

As with SL, informed customers know that the cost per unit dramatically drops when you order multiple quantities of different parts. In traditional pricing, the first widget costs $300, so traditional quoting would estimate that 10 different widgets would cost $3,000. However, with family builds, the first widget costs $300 because the operational overhead is factored into the first part, but the next 9 are made without a factor of operational waste, so the total price of 10 would be $1,500. Buyers like seeing the cost per unit drop from $300 to $150 each. With family build, everybody wins. The savings are considerable.

## • Multiples of One Part

Experienced customers also know that the cost per unit dramatically drops when ordering multiple quantities of one part. In traditional pricing, a battery door costs $200, so 3 battery doors would cost $600. However, using family build thinking, the first battery door costs $200, and the next 2 are made without operational waste. Therefore, the total price of 3 battery doors is $350.

## • Single Material

Significant savings result from building all of your SLS parts in a single material, combined with a family build. Every time the SLS machine stops to change materials, extra labor and operational overhead are factored into the order. Plan carefully to find one material that suits all your needs.

## How Do I Waste Money in the SLS Environment?

If you are really committed to wasting money, be sure to ask for the wrong part orientation, order parts one at a time, and use several materials to increase the operational overhead factor. You can also send bad or outdated CAD or STL data, and haggle using the old-fashioned manual quoting methods. Service providers will avoid you at all costs.

Joking aside, cutting and rejoining big parts in SLS is a real challenge that can waste money. The SLS process often presents tolerance issues resulting from material warpage and heat, making splitting and rejoining a part very tricky. Twist and distortion in a part are usually not visible until it's too late. Design thick areas that can easily be rejoined, or consider using another process.

## How Do I Save Time Using SLS?

Because parts are built in a vat of powder without support structures, SLS parts require less finishing. While an SL part takes hours to finish, SLS is a batch finishing process where all the parts are dug out of the powder at once. The only time-eater here is that SLS powder takes hours to cool down.

SLS is unique in its capacity to build nested parts, like the hollow, egg-shaped Matrioshka dolls of Russia that have progressively smaller versions inside. If that's hard to visualize, imagine how a chef makes a Thanksgiving "Turducken." He stuffs a chicken inside a duck inside a turkey. With SLS you can also build a box inside a box inside a box. Therefore, you can use this process to build functional parts that fit together. So, you can build fully functional assemblies with SLS since the supports are not in the way. If you need a meshing gear and housing, the SLS system can build rotating gears inside a housing. Moving parts make great samples for your customers. They love to see and play with actual working parts.

Another huge timesaver is using layered manufacturing to build parts for production. Companies needing a low-volume run—only 100 to 30,000 parts—use SLS to completely bypass the hassles and expense of tooling.

As with all AF processes, be sure you learn as much as possible about materials before launching an SLS project. Always double-check your CAD files for faceting, and be sure to send the correct version or latest CAD data to your service provider. Get comfortable using instant quoting and save valuable engineering hours by eliminating the laborious, manual quoting process.

### How Do I Waste Time in the SLS Environment?

Typical time-eaters lurk in every process. Informed customers understand the pitfalls of great expectations based on wrong assumptions. If you select the wrong material based on your part's geometry, you can get a warped surprise. If you need flat, thin parts or fine features, SLS is not the process for you. **Be sure to understand the limitations inherent in any process you choose.**

### ▓ The Keys to a Brand New Maserati

By now you're revved up about SLS technology and the functional, durable, heat-resistant parts it offers. You've witnessed Johnny Quickparts as he saved Overrun's head with an ingenious design and a low-volume production run in SLS. You've learned that SLS builds in a fully 3D envelope, and makes macho parts out of thermoplastic powder. You know the most common mistakes in the SLS process result from sending wrong CAD files and designing out of tolerance. Wrong orientation and poor communication can make an SLS project flop.

You also know that SLS has no support structures and that its tolerance is less than SL. By now you are loving economies of scale and you want to apply it to everything, especially pizzas. You know how to avoid the SLS pitfalls that can happen with flat parts or fine features, and you are one step closer to becoming QuickSMART.

As Johnny would say,

**"Fire it up."**

# Chapter 4

# Fused Deposition Modeling

## TWO NOZZLES ARE BETTER THAN ONE

*"Sometimes turtles win the race."*

## QuickSMART

**D**efinition: Fused Deposition Modeling (FDM) is the strongest, but slowest, of the major solid-based AF systems that produce plastic parts from cross-sectioned CAD models. Electronic CAD design data is converted to Standard Tessellation Language (STL) format, and then special software slices the CAD model into thin layers and creates build instructions for the machine. A heated head with two extrusion nozzles builds the part, layer by layer, pressing spools of filament through an industrial "hot glue gun." FDM is a unique two-material process that provides major strength to parts. The first nozzle dispenses melted support material that dissolves away in water; the second nozzle extrudes the permanent base material. A plastic physical model is made of many micro-layers of melted filament that solidifies immediately upon cooling.

**Why You Need It**: To produce fully functional parts; to test form, fit, and function with near-production quality parts resistant to heat, water, and chemicals; to utilize the good mechanical strength of FDM materials; it has approximately 80% of the strength of injection-molded Acrylonitrile Butadiene Styrene (ABS); and it is the only process that offers a variety of color choices.

**Ideal Uses**: Conceptual and engineering models; patterns and masters for tooling; fully functional prototypes for design, analysis, and testing; durable, closest to real production parts; and an excellent choice for vacuum forming tools.

## ▓ FDM Background

Like the other AF processes FDM works on an additive principle by depositing material in layers. Similar to SL, an FDM part is also built from the ground up, in layers, as hot plastic filament hardens immediately after extrusion from the dispensing nozzle. Surface chemistry, thermal energy, and layered manufacturing are the basis of FDM technology.

FDM was developed by Scott Crump in 1988 and was commercialized by Stratasys, Inc., in 1990. The FDM equipment and materials are marketed exclusively by Stratasys in Eden Prairie, Minnesota. Stratasys also makes the Dimension 3D printer, an office version based on FDM technology, which has sold thousands worldwide. Relatively speaking, very few FDM systems are sold; however, Stratasys leads the equipment manufacturers of the industry in total units sold, when combining FDM systems and Dimension printers. System sales for FDM have been especially brisk in Asia, although clones of FDM systems could become a major competitor overseas.

> **QuickTip:** Because of water-soluble supports, complex or internal geometries are possible with FDM.

Unique among the "big boy" additive processes of SL and SLS, FDM uses bi-material deposition, which requires a base material for the actual part and a water-soluble material for creating temporary support structures. The two-material process is unique to Stratasys, as is the water-soluble material that dissolves away from the base material when immersed in a heated salt bath and blasted with ultrasonic waves. **This material innovation cleverly does away with costly post-processing time and the risk of part damage during cleanup.** The plastic filament is sold on a spool much like a roll of Weed Eater wire. The recipe for these expensive materials is, of course, top secret.

Early on, Stratasys demonstrated pure marketing genius by naming its own materials with the common engineering names of standard plastics, such as ABS and Polycarbonate (PC). The FDM filament is technically ABS-like, not pure ABS. An informed engineer knows that there are many types of ABS available and will shop carefully.

By owning the commonly-known, generic plastics names, Stratasys was able to stage a massive marketing coup by extending its faux material into the real material world. As the market has shifted from "early adopters" to the "early majority," users assume that all materials are as named. Therefore, Stratasys has influenced the perception of engineers to think they are getting ABS, the same material as the end-use finished product. By owning the real name of the end-use plastic, Stratasys "owns" the minds of many engineers. In marketing, where perception is everything, owning a mind is pure gold.

> **QuickTip:** If you are one of the billions of people who can't say "Acrylonitrile Butadiene Styrene" it's ok to say ABS. It's a generic end-use engineering plastic.

If you compare an injection-molded part made in standard ABS to an FDM part made in Stratasys ABS, the injection-molded part will show that it was pressurized in a closed volume, packing the molecules tightly together. The FDM process does not pressurize the molecules, so the part is lighter, more conceptual, and slightly toy like.

The good news is that **FDM parts are durable and functional.** In fact, many FDM parts can be used in real working environments. The bad news is that **the FDM process is very slow,** which drives up the cost of parts.

## ⋮⋮ FDM Process—A Big Hot Glue Gun

Engineering funny-guys sometimes refer to the FDM system as a "big hot glue gun," the kind grandmothers use for making indestructible Christmas doodads and Easter baskets. An apt comparison, FDM converts virtual models into real parts by laying down many layers of a hot, glue-like plastic filament. The dispensing nozzle is heated to melt the material then moved in both horizontal and vertical directions by a numerically controlled mechanism, controlled by CAD software. The FDM process is very safe in that materials are non-toxic and there is no exposure to a laser. Like SL and SLS, the FDM process is classified as additive fabrication or layered manufacturing, names that describe the process of producing a physical part with successive layers.

The FDM system consists of a CNC-controlled table with an x-y build platform, foam base, liquefier head, two material spools, extrusion tips, and a controlling system. Similar to the SL and SLS processes, a CAD model also provides cross-sectioned build information to the FDM system. Layering technology is performed through proprietary software, slicing the CAD data into layers. These layers become build instructions in the FDM process.

### Step by Step with FDM

First, the operator uses software to slice a 3D CAD model into thin layers in the z-axis. This data drives the extrusion head of the FDM system. Plastic filament on a spool is fed through a heated extrusion nozzle. The machine traces out the cross section of each layer, laying down a continuous stream of molten material that cools almost instantly. The second filament is fed from an adjacent nozzle for support material, which is used to support undercut or overhanging features. After the entire cross section is outlined with melted material, the build platform descends by one layer thickness to make room for the next layer. The layering process repeats until the 3D part is fully built. The temperature of the build chamber is precisely controlled to remain slightly below the materials' melting point so that only a little extra heat is required to melt the filament.

## :: FDM Applications—Stop and Smell the Glue

Look around you. Signs of FDM are everywhere. The door handle assembly of your car was tested in prototype stage using FDM. The commercial airplane window frame, also prototyped in FDM, was first tested at 35,000 feet to make sure that all passengers would remain safely inside the plane. Components in your city's fire truck engines were vibration-tested using FDM. Your neighbor's Hyundai was designed using FDM to ensure dimensional accuracy and stability. A number of race cars on the major speedways have aerodynamically tested components using FDM prototypes.

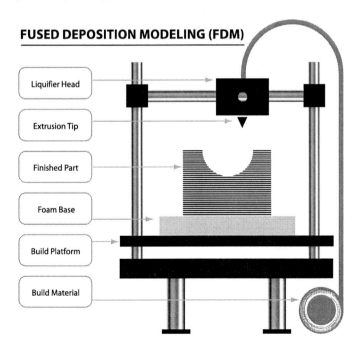

**FUSED DEPOSITION MODELING (FDM)**

Liquifier Head

Extrusion Tip

Finished Part

Foam Base

Build Platform

Build Material

Once you understand it, AF makes you look at the world in a new way. You start seeing shapes, materials, and processes, things you never really notice as a hurried shopper. **By applying AF, engineers can radically reduce product development time.** Product developers are now able to enjoy creating more innovative products—the cool, sleek, and sexy—at prices the consumer can afford. The latest version of the world we live in is a result of this virtual-to-reality evolution. A huge player in compressing time cycles, FDM produces quick physical replicas and actual parts. As such, FDM has an opportunity to help lead the trend toward low-volume production.

### Industry Overview—A Nerd's Eye View of FDM

FDM is a sound technology with seemingly nowhere to grow. Innovations in machine design and materials are long overdue. This process is like a contemporary dinosaur, slowly grazing the green grass of the new paradigm. However, it's still thriving, thanks to its excellent material properties that no other process can touch. Perhaps it is already approaching its own extinction, but **FDM is still the best process for creating the most real production parts without tooling.** Unfortunately, the slow pace of the process is a real hindrance to designers on a deadline. The speed of the process also makes it hard for service providers to support the FDM process. Currently, there are only a few service providers that provide FDM parts in the world.

Because FDM parts can take weeks to build, service providers can't turn a profit on a slow process that builds only one part at a time on a 2D build platform. Original equipment manufacturers (OEMs) with their own prototyping labs can afford to purchase FDM equipment; they keep it running 24/7 to get the most value out of it. And yet, Stratasys still leads AF in equipment sales thanks to its 3D printer, Dimension, a stripped-down FDM system. Last year, thousands of engineers and designers bought Dimension for concept "printing" of 3D objects directly from CAD.

The difference between the FDM system and the Dimension printer specifications is considerable, while the materials they use are essentially the same. The Dimension—good for concept verification—uses a bigger tip and provides less accuracy, while the FDM system makes functional, full-strength parts with higher accuracy.

A committed leader in the field of additive fabrication, Stratasys, is the exclusive provider of FDM technology in the world, except for the problem of fast-paced cloning in China. Stratasys has been a consistent force in driving the use of FDM parts worldwide and is now leading the trend for layered manufacturing. Promoting the use of AF parts for volumes of 100 to 5,000 parts per run, **layered manufacturing allows parts to be made without tooling, saving considerable time and money.** Stratasys has become a world-class leader through solid performance and marketing.

FDM was critical in introducing the concept of layered manufacturing to the industrial world. In the same way short-lived electronic typewriters were an "in-between technology" that quickly gave way to far superior PCs, FDM has introduced a revolutionary foundation for building low-volume parts without tooling. Because FDM is much slower than SLS, the FDM cost per build will price itself out of existence. While it seems that FDM technology is at a dead-end for improvements, it is still making a significant contribution to industrial developments.

### *Low-Volume Layered Manufacturing*

Low-Volume Layered Manufacturing (LVLM) is a very powerful, evolving trend that continues to merge the worlds of engineering and manufacturing. The LVLM approach extends the use of the additive fabricated part as the actual end-use part for the product. Since this trend is very new, many names have been associated with it, including rapid manufacturing and Direct

Digital Manufacturing (DDM), as recently decided by the Society of Manufacturing Engineers (SME). Since these names are not descriptive enough, LVLM seems appropriate. Regardless of the name, the trend is that **product developers are using traditional AF processes to produce end-use parts.**

The LVLM approach has been the dream of many pioneers since the early days of AF. The capacity to make end-use parts this way would give these dreamers and doers the acceptance and credibility they longed for in the manufacturing world. These pioneers have wanted to be equal with the injection-molding technologies and have a place at the "adult table." More and more, manufacturing of the twenty-first century is taking advantage of the power of AF and efficiently producing end-use parts. Since this is a dynamic technology environment, imagine these scenarios to help you understand the significant advantages of LVLM and how it can be used.

Imagine the flexibility to design a part for its purpose without regard to the many constraints that are imposed by traditional manufacturing. Wouldn't it be nice to not give a damn about draft? How about eliminating the need to have any tooling to manufacture the part? This would be a huge savings in time and money. Without tooling, parts could be manufactured as they are needed, thus reducing inventory waste and allowing for design changes to be incorporated quickly and economically.

With this new-found design flexibility, you could also consolidate the many parts of your product into more complex and useful parts. This would reduce the parts in your product, simplify your bill of materials, and make the product work better. LVLM helps the designers of the world who are experts at driving CAD but weak at designing for manufacturability. Check out the next image to see how an assembly of some many parts is consolidated into two parts. It doesn't get much better than that!

## DESIGN FLEXIBILITY

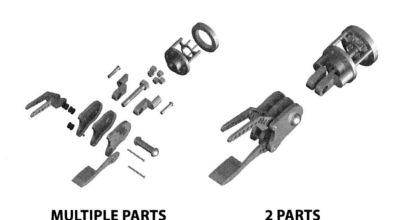

**MULTIPLE PARTS**                     **2 PARTS**

So, Alice, what's wrong with this Wonderland? As with all paradigm shifts, there are the seen and unforeseen challenges. The industry is still learning how and when to apply AF for end-use parts. Currently, one of the biggest obstacles seems to be that the parts from these layered manufacturing systems simply are not injection-molded parts. Duh! This puzzlingly obvious statement reveals a cautionary truth: LVLM parts don't look, feel, or, in some cases, act like their injection-molded cousins. Also, LVLM presents a challenge when secondary finishing is needed. Of course, with flexibility there is accountability, so the ability to easily make design changes also requires the ability to manage design versions so that you will know what part is actually being used in the world. Lastly, the accepted quality control methods still need to be adapted to handle the variability that can be introduced in the part. At the end of the day, a part is not a part is not a part.

To really drive the evolution of LVLM, the manufacturers of additive systems need to take a focused role in their development programs to address the inherent shortcomings of their processes. The easy-to-identify actions for Stratasys, 3D Systems, and any newcomers include the development of quality end-use materials, manufacturing processes that emulate injection-molded parts, and further education about how to better apply these approaches to product development strategies. These companies must decide whether they are developing systems to make prototypes or to manufacture parts.

So, when do you, the product developer, use this powerful LVLM approach? Currently, it seems that the best applications are for internal parts that will never be seen or heard. Since the better LVLM processes make some ugly parts, it is better if we don't have to know what the parts look like. Also, non-critical parts are good LVLM candidates, as well as parts that can accept variability in their tolerances. Some of the best uses come from situations where you need to get your product to market and don't have time to wait for tooling or you are not ready to finalize the design. The LVLM approach will provide you an option to get parts quickly and economically while you continue to make progress in your design.

What does "economical" really mean? With LVLM, you will need to be sitting down when you get your first quote for 100 units. The part price will be a shocker! Instead of low pricing in cents or dollars, your pricing will be in the tens or hundreds of dollars each. The good news is that you don't have to spend $30,000 for tooling and you still get all of the other benefits as well. Depending on your product or need, LVLM can be an economical solution for your project.

As you continue to read, you will discover Low-Volume Injection Molding (LVIM), part of the Product Developer's Toolbox. LVLM could have been included in that discussion as well. However, since the core of LVLM is additive fabrication, its proper introduction was within the FDM context. While FDM and SLS are both valid processes to use in this approach, Stratasys has led the way in providing materials and education to the industry to make LVLM a viable product development tool.

## *Resolution—The Nitty Gritty*

Resolution definitely matters with FDM, but it differs from the other additive processes. Think of FDM as a marker pen with three different tip sizes, much like your prized stash of Sharpies: fine, medium, and bold. FDM works with variables of layer thickness—as determined by CAD slices—and resolutions determined by these tip sizes, including 0.010 inch (ten thousandths), 0.007 inch (seven thousandths), and 0.005 inch (five thousandths). Standard layer thickness is 0.01 inch (one thousandth), and minimum wall thickness is 0.02 inch (two thousandths).

Like a pen point, the FDM tip thickness affects the fineness of features it can create. Imagine that you are drawing an architectural ornament that features a "bas-relief" of a tree. For the pen sketch, you would use a fine tip to detail the small leaves and a thick tip to color in the wide trunk area. When building this same relief design with FDM, you only get one tip, so choose wisely. Consider the tradeoffs. The thinner the extrusion tip, the longer the build time. The thicker the tip, the less detail that can be rendered.

Discuss tip size with your service provider before you start because the FDM process is limited to one tip per build. You can't switch tips in the middle of the process. Evaluate tradeoffs to preserve the most important features.

> **QuickTip:** "Bas-relief" (pronounced: bah-releef) is a subtractive method of sculpting that carves away select surface areas of a flat piece of stone or metal, resulting in a raised pattern. Bas-relief is often used in architectural detail.

## Benefits—The Buzz is Real

A key difference between FDM and the other AF processes is that FDM materials (ABS and PC), allow for the production of durable parts that act like real parts. Parts made with FDM are very strong and have a long shelf life. It's the only non-tooled process that is very close to imitating "real parts" made with tooling.

The diversity of FDM applications is impressive. Most AF parts are used as communication tools for engineers and toolmakers, test models, studies, or patterns. FDM leaps directly from CAD into what some call rapid manufacturing, DDM, or LVLM, which makes the parts without the tooling. The FDM process can also be used for vacuum forming tools. Since FDM parts are porous, air is sucked through them easily, making good tools. However, unknown variables in heat can cause issues.

Case studies from trade journals demonstrate profound cost and time savings with FDM: A window manufacturer cut costs by a whopping 85% in its conceptual design phase of a snap fit. An automotive contractor cut tooling cost and time by 50% by using FDM for its design for fenders and bumpers. In creating auditorium seating, a plastics company saved over $50,000 and eight weeks using FDM. Another automotive giant used FDM to manufacture jigs used in the assembly of automobiles. Theme parks have used FDM to create durable parts for ticket reservation equipment. A satellite launcher project made use of FDM to produce rocket stage canisters. World-class surgeons and biomedical designers who know about FDM often enhance and sometimes even save lives with cranial plates and precisely fitting prosthetic limbs.

### A Very Tall Tale—Johnny Quickparts and the Eccentric Texan Challenge

So far that week, Acme Design Corporation was operating swimmingly well. Not a single crisis had reared its ugly head. Either business was fatally slow or Acme management was inadvertently doing something right. But even by Johnny's most objective estimation, Acme's "Bring-Grandma-to-Work Day" was not the best corporate outreach program his management had ever dreamed up. The event usually started an annual Casserole War, which exploded just before the No-Fat Dessert Winner was announced.

With all the evolution happening in rapid technologies, Johnny pondered a universal paradigm shift as he studied the youngest-looking Grannies he had ever seen—except for his grandma, Granny Quickparts, who looked like a traditional flannel nightgown type. Johnny's designing mind did notice that this year in particular, most of these silver citizens didn't look like Grandmas at all.

Johnny deduced that something good but strange had infiltrated everything and everyone, not just product development. Technology was somehow touching these matriarchs. No longer blue-haired grannies, these mature cheerleaders looked engineered backward toward eternal youth, or at the very least, a sprightly age 45. "Weird but true," Johnny said out loud, as he gulped down a puzzling tofu and whipped cream extravaganza. Even the fat-free desserts tasted suspiciously altered, as if a plastics engineer was in the kitchen. Looking around the room, Johnny momentarily lost his head over a spunky 75-year-old who looked not a day older than a flawless, reverse-engineered 62. Johnny hit the pause button on his fantasy when Bob Overrun interrupted the luncheon and pulled Johnny upstairs by his belt.

Acme's best-paying customer, the wayward Texan billionaire inventor, Mr. Ima Kook, was on a rampage. Bored with his huge ranch and tumbleweeds, he suddenly needed 100 durable, plastic Texas-shaped swimming pool filters. He could not wait for tooling to be built and wanted them in seven days flat. His filters had to be not only water-resistant but impervious to chemicals like chlorine and children's pee. Mr. Kook was dismayed with the previous filter. It was leaky and warped and did not screen out gunk they way it should.

It was obvious to Johnny that a Texan ego had once again overruled the designer's common sense. As Johnny turned the Texas-shaped filter over and over in his hands, the Lone Star State looked like a large, dumb bird flying upside-down. Johnny had never traveled outside of his hometown, Huntsville, Alabama, the birthplace of rockets and the highest number of engineers per capita in the world. But he still felt that the "bigness" of Texas should be upheld by anyone living below the Mason-Dixon Line. Once he sorted out his odd loyalty to Texas, Johnny could tell that the filter had failed due to a lousy design and flimsy material. He knew there was a way to make it stay hard in a continuously wet world.

Since no job was too weird for Johnny, he fully surrendered to dignifying the task at hand. He downloaded the Texas state song, "Yellow Rose of Texas," from the Internet. Then he hauled himself up by the python bootstraps. Like any good Texan, he checked for his ten-gallon hat, his six-shooter, his RC Cola, and Moon Pies. Johnny's heroic heart began swelling with new vision. His filter design would have more body, more muscle, more *huevos!* Johnny's filter would snap into place so tight that it would sound like a gunshot at high noon. He searched his right brain until he found a special color for his filter plastic. A big clue was the state song. Texans love yellow; they even take credit for inventing it...yellow roses, yellow birds, and Old Yeller. It took Johnny 30 minutes to redesign the Lone Star filter with specially beveled walls to ensure a perfect fit. Surely Mr. Kook would be happy that Johnny was eliminating the need for expensive tooling.

Thanks to hours of reading in bed as a single guy, Johnny knew that FDM was the only way to go. For smaller parts like this, the process would be fast enough.

**The FDM filters would be durable, waterproof, and chemically resistant.**

Besides, the ABS material came in bright yellow. But Johnny was not ready to deal with his feelings ignited by his favorite sales rep, Helen Helpalot. He chose to upload his file for an instant quote all by himself. It was so easy to log in, upload, and buy. Done. He didn't even blush.

In less than five seconds, his phone rang anyway. Helen was calling to thank him for his order. She even congratulated him for a highly creative solution with FDM. Then Helen directed her high beams on Johnny's heart, "You've got a one-of-a-kind mind, Big Guy, and you'll have a hundred FDM filters, bright yellow, in three days." Johnny's head was spinning inside the closed universe. Helen's sincerity made his inner man-self quickly expand beyond his capacity to comprehend, so he made it brief: "Helen, you rock!" After the call, Johnny steadied himself at the water fountain. "It's a mathematical probability that maybe I really am lovable," he muttered aloud.

When the yellow FDM filters were delivered, Mr. Kook slapped Johnny on the back and made a frightening hog call, indicating total customer approval. When Mr. Overrun asked the customer to evaluate Johnny's performance, Mr. Kook said emphatically, "He's got it by the tail going downhill!"

Mr. Overrun was starting to acknowledge Johnny's genius even though it made him twitch. He would now revise Johnny's annual performance evaluation, raising him from "Mediocre" to "Fair." Overrun carefully rationalized his caution: sure, Johnny could generate good ideas, get things done, and save the day, alright, but could he do it consistently?

## ▦ Boring but Necessary—Understanding FDM Limitations

Knowing the inherent limitations of each process is essential to your success in the product development field. Good designers always intend "robust designs,"—designs that are easy to understand and change. With layered manufacturing now available, your job description just rapidly expanded. Be sure to take the hidden blinders off your critical thinking to make your results even better. You now need to know not only the latest design tools but also the latest processes and materials. Learn the real versus the wished-for capacity of each.

## Part Features

**Large, bulky parts work best with FDM.** FDM is typically better for parts with larger features—greater than 0.010 inch (ten thousandths)—since extrusion tip size affects layer thickness. Features less than 0.30 inch (thirty thousandths) are high-risk and should be produced using SL or SLS.

## Tolerances and Accuracy—Real World

Dimensional tolerances for FDM are ± 0.005 inch (five thousandths) for the first inch, and ± 0.002 inch (two thousandths) for each additional inch. In the z height (vertical), standard tolerances of ± 0.01 inch (one thousandth) for the first inch, ± 0.002 inch (two thousandths) on every inch thereafter. The standard layer thickness is 0.01 inch (one ten-thousandth), and minimum wall thickness is 0.02 inch (twenty thousandths).

Informed engineers know that the CNC driving this process is highly accurate; however, the extrusion process can affect the accuracy of the part when the melted material solidifies. Commonly accommodated by the operator, shrink factor can also affect part accuracy.

## Part Orientation—How to "Grok" XYZ

The key differentiator between FDM and other additive processes is flexibility. Unlike SL and SLS, the **build time and quality of FDM parts are not affected by part orientation.** In SL and SLS, the horizontal orientation produces faster, better cylindrical parts. Horizontal or vertical, the FDM build takes the same amount of time, costs the same, and renders the same quality.

Resolution does not matter in relation to FDM build time or cost, but it always matters with respect to features. If a part has more than one critical feature, an informed engineer using FDM lets the service provider know which feature is most important to him so that orientation will be planned accordingly. If you don't do this, your part's feature definitions—overhangs and undercuts—could suffer.

### *Size Matters—Are We Surprised?*

The FDM build envelope size is similar to the SL build envelope. With FDM there are several sizes available. The standard envelope is 10 x 10 x 10 inches. The medium envelope is 14 x 16 x 16 (for polycarbonate builds). The largest envelope is approximately 23 x 19 x 23 inches (for ABS builds). Like the SL machine, FDM is also limited to a 2D build platform, thereby limiting the number of parts that can be built at one time. However, there is no need to limit part size. Parts bigger than the specified envelope can be built in pieces then rejoined together.

### *Materials are a Nightmare*

As mentioned previously, Stratasys named its own materials with names of standard plastics, such as ABS. FDM filaments are technically ABS-like, not pure ABS, or PC. Don't make the mistake of thinking that you are designing FDM parts in pure ABS plastic, the same material as the end-use finished product. It's very important to know the difference.

### • Material Types—It's All "Butter"

Asking Stratasys for ABS is like asking someone at the dinner table to "Please, pass the butter." No matter what yellow substance they hand you, you will use it, even if it's margarine or any of its cousins.

The reason it's important to know the difference between true, standard ABS and the Stratasys ABS-like material is that there are differing inherent characteristics in each material that react in a variety of ways based on part geometry and heat. Choose the material that will support the functionality of your design. If you compare samples, you will immediately understand, feel, and know that there is no substitute for injection-molded ABS.

Several other materials can be used in the FDM process, such as polyphenylsulfones (PPSF) or polyester, each with unique tradeoffs in strength and temperature.

The only additive material that comes in a variety of colors, including black, white, red, green, blue, grey, and yellow, is ABS.

### • Water-Resistant—Glub, Glub

Sanding isn't possible with ABS because it's too tough. Early in development, Stratasys realized that post-processing tough plastic parts would be a nightmare if the FDM process required support structures to be removed and sanded away. Instead, they developed a two-material system to avoid these problems. The first base material, a very tough plastic dedicated to the part itself, is impervious to water and chemicals. The second material is used as temporary support during the build. It dissolves away easily after the build when immersed in a warm, soapy bath. Stratasys may be the only company to offer this unique two-material process that results in quick, easy cleanup.

### Temperature—FDM Likes it Hot

Like SLS, FDM parts are born of heat and can withstand high-heat environments of over 200ºF. Therefore, **FDM is an excellent choice for heavy industrial applications found in aerospace and automotive applications.**

### How Will My Part Look?

Beauty is in the eye of the tool holder, right? FDM parts are not especially gorgeous, but it depends on what you need. The finish is somewhat granular and rough to the touch. Too tough for sanding, FDM parts can feature color options to make them more aesthetically pleasing.

• **Surface and Post-Processing**
There are no special finishes available for these super-tough parts, and sanding is not an option. Thanks to the water-soluble support system, FDM parts don't need additional post-processing. Paint is not recommended for the somewhat porous FDM parts that are made with a built-in, superfine cross-hatching pattern (the extrusion head travels in vertical-horizontal paths). The more experienced FDM user knows that "WYGIWYW"—What you get is what you want.

### How Long Does it Take to Get My Part?
The standard turnaround time for small FDM parts is three to five days. However, a large part that is 100 cubic inches in volume can take two to three weeks just to build. Even with these long build times, FDM can still save you tons of money by allowing you to make design changes early in the design process while it's still relatively inexpensive. This slow process makes some engineers rethink the term "rapid" prototyping!

## Saving Time, Saving Money—The Rules are Different with FDM

### How Do I Save Money Using FDM?
Remember all those cost savings secrets you learned with SL and SLS? As television's Tony Soprano would say, "Fagetta 'bout it!" FDM offers very few cost saving options. Economies of scale and the family build concept do not apply to FDM. Part orientation matters not to the build time because each FDM part contributes equally to the build time. Building in a single material offers no savings either because it takes as long as it takes.

The only way to save money using FDM is to buy the cheapest parts on the market. But let the buyer beware! By now you understand that the FDM process is more an art than a science. This means that not all service providers will produce suitable, quality parts. Many service providers do not have the latest equipment or upgrades. Also, their parts may lack quality because they may not use Stratasys-certified materials. Smaller service providers often sell parts made on 3D printers and call them "FDM parts." In cases like this, if you are expecting authentic FDM parts, you will be very disappointed. **Dimension parts are good for concept visualization, but they are not the same as FDM parts.** Dimension parts feel lighter and will not function as real parts. It's worth a look and feel to really know the difference between the two. FDM parts are real parts. How do you tell? If two FDM parts interconnect, they can function consistently and hold a high tolerance.

**QuickTip:** Small service providers often misrepresent 3D printer-made parts, selling them as authentic FDM. Wipe the milk off your moustache and ask, "Got real parts?"

## How Do I Waste Money in the FDM Environment?

Wrong orientation happens when a customer fails to communicate critical features to his or her service provider. This results in an expensive rework. Be sure to let your service provider know to ensure that the operator will protect those features with proper orientation. Remember that orientation for FDM is feature dependent. Plan ahead to ship groups of parts together. Avoid wasting loads of money in shipping single parts.

## How Do I Save Time Using FDM?

Build failures are obviously devastating on large parts with slow build times. A power outage can make even the most macho engineering manager cry buckets of tears. Be sure to double-check your STL faceting. The FDM process requires triangles or polygons in the STL file to be nearly perfect; otherwise, a part can fail. Because it is hard to detect bad STL data during the build, safeguard your project. Ask your FDM-qualified customer service representative to check your file with STL viewing software before building.

As always, make sure you send the correct CAD file, and use instant quoting to save your valuable engineering man-hours.

## How Do I Waste Time in the FDM Environment?

If you really want to waste time in the FDM world, be sure to select wrong material for process, send the wrong version of CAD data or a bad STL file, and use old-fashioned manual quoting! You could also forget what unit of measure you are using. Countless jobs have been lost to confusion about inches versus millimeters! (Remember NASA's Hubble renovation?) You will definitely waste time if you expect the FDM process to do small features well—it won't. Engineers often overlook this limitation when designing snap fits in FDM.

### ⁘ The Keys to a Brand New Bentley

By now you know that FDM is like a two-headed turtle that produces parts very close to real, ABS-production parts. You could explain to your buddies why "ABS" is ABS, but not really ABS! You've been in the trenches with Johnny Quickparts as he dazzled an impossible Texan with FDM, chosen for its excellent material properties of durability and chemical resistance. You know that FDM orientation is feature dependent. You've been cautioned about the difference between 3D-printed parts and real FDM parts, often marketed as the same thing. Most importantly, you know that you can completely eliminate tooling using FDM.

As Johnny would say,

### "Let's test this thing!"

# Chapter 5

# Low-Tech Ways to Make Parts

## OLDIES BUT GOODIES

*"If all else fails, go back to the basics."*

## ⁘ Two Industry Basics

The first three processes of this book, Stereolithography (SL), Selective Laser Sintering (SLS), and Fused Deposition Modeling (FDM), are considered additive fabrication (AF) in the manufacturing industry and the latest in the technical evolution of manufacturing. Cast Urethane (CU) is classified as a formative fabrication process, while Computer Numerically Controlled (CNC) machining is classified as subtractive fabrication. While defined here as low-tech, it is all relative since there are some very sophisticated and advanced CNC technologies. Obviously, there is more than one way to make a part. Sometimes the "faster, cheaper, better" way to get your parts made is using low-tech or even no-tech tools. Think back to first grade when you made your handprint in a pie pan of wet plaster. Any other process would have been overkill.

This chapter highlights two standard ways to make plastic parts: CU and CNC machining. Both have been the backbone of the manufacturing industry for over 60 years. If you don't need advanced technology, go back to the basics. These foundational processes can provide an important, low-cost transition between the phases of prototyping and tooling. Most people use these two methods to get a look-see-touch on a prototype before investing further in the iterative design process.

The CU process would typically be needed midway in the product development cycle to test marketing, packaging, and form, fit, and function. A typical CU run is about 10 to 50 parts. After parts are cast in urethane, typically design improvements are made. Based on those changes, the improved, low-volume tool or the production tool is then made.

## ▓ Cast Urethane Parts from RTV Tooling

### *QuickSMART*

**Definition**: CU parts are durable, real-looking parts made from a two-part polyurethane material and formed in a silicone mold under relatively low pressure. The CU process has three distinct steps: pattern, room temperature vulcanization (RTV) tool, and part. The RTV tool is made from an impression of the master pattern in rubbery material, leaving a negative space to fill. The plastic part is made by pouring polyurethane into the negative space of the RTV tool. When the part is removed from the tool, it is a positive identical to the master pattern.

**Why You Need It**: To get durable, representative models quickly; to produce tooling and parts much faster than with production tools; to get real-looking, aesthetic parts with color, texture, and finish like production parts; to save money with a less expensive process than tooled parts; in some cases, to get functionally operating parts; and to validate product-to-market before committing to tooling.

**Ideal Uses**: Tradeshow or marketing samples; fit, form, function with the complete assembly; functional evaluations; test packaging design; conceptual models; pre-production parts; very low-volume production; rapid and inexpensive test assemblies; fixtures for production processes; to preserve a master part; to keep a master or "go-by" ready; use in reverse engineering to make a pattern of a part that can be destroyed; and get a few parts to setup a production or packing line.

### Cast Urethane's Evolution

Plastic polymers were named as such in 1937, when Otto Bayer and his team of chemists and inventors developed a polymer process related to spinnable products. Shortly thereafter he patented a unique kind of polymer called polyurethane. Polyurethane soon became synonymous with a highly versatile class of plastic made famous in the 1967 movie, *The Graduate*.

There are many highly complex facets to the chemistry behind the seemingly simple process of CU technology. For our purposes, the basic CU process brings together two or more liquid polyurethane chemicals to cause the chemical reaction of "phasing," or changing liquid to a solid state. The ability to "gel" liquids into solids is what allows urethane to fill a silicone mold with almost any shape imaginable and hold it. Because of form-friendliness, part toughness, and durability, **CU is widely used across all industries and product types.** CU represents a quick way to get real plastic parts by copying a pattern. Thanks to AF processes that make quick tool patterns, CU has become even more prevalent in the last 15 years. While CU parts are not, technically speaking, true, engineered plastic parts, they "behave" closely enough to be useful for many kinds of testing.

CU parts can have aesthetic properties like color, texture, and finish that look just like the more expensive production parts. This is due to the properties of materials used in CU that are very similar to production thermoplastic but are not exactly the same. Urethane parts can operate functionally in many cases, and they are less expensive than tooled parts. CU tooling is much faster than having production tooled parts made, so parts can start being produced faster. CU parts can be produced anywhere from several minutes to a day, depending on the material.

> **QuickTip:** CU is an excellent method to get parts for marketing literature and packaging studies.

Disadvantages of the CU process can be a deterrent for those needing high tolerances or production quality. Parts cast from RTV tooling can only look as good as the pattern used. In other words, a bad or poor pattern will replicate poor parts and copy any problems in the pattern's feature definition, tolerances, or aesthetics. Mechanical properties of materials in CU are much less than thermoplastic properties, particularly in regard to temperature, impact, strength, and flexibility.

Thermoplastic materials cannot be used in the RTV mold because RTV molds are not conducive to the high temperature and high pressure required to shape thermoplastics. Moreover, the quality of the tool and part will degrade with each use. On average, an RTV tool wears out quickly, typically producing an average of 25 parts. Simple tools can run up to 50 parts per tool, whereas complex parts may only get 5 parts per tool. It is very common to produce bad trial parts with several uses of a tool. Since a tool gets 5 to 50 uses or shots each, bad parts can waste shots and valuable production time.

## ▪▪ CU Process

When making a CU part, there are three distinct steps in the process. Just remember "pattern, tool, part" and you will navigate your way successfully through this process.

The rubbery RTV tool or mold used in the CU process is made of a kind of silicone, flexible enough, when cured, to remove the part but strong enough to keep its shape. RTV silicone is "castable" around any pattern in a room temperature environment.

The CU mold must be made from a physical pattern, typically an SL part because it provides the best process for a smooth finish. For reverse engineering purposes, you can use any physical part in inventory. The key to successfully casting plastic parts is to use a mold that is flexible enough to remove the part but strong enough to be reused. RTV tools are the fastest, most accurate, and least expensive way to duplicate directly off patterns in very low quantities of 1 to 10 or more. These tools will duplicate details and textures present on the original part or pattern.

The master pattern is usually an RP model and can be made from any AF process described earlier: SL, SLS, FDM, CNC, or an actual part. SL is the most commonly used process. The pattern is hand-finished to a high level, and texturing is added if needed. If inserts are required, the insert is added to the pattern prior to casting the RTV tool.

Careful consideration is given to large parts and certain casting materials to compensate for shrinkage, as degrees of shrinkage will occur each time in a three-part replication process. When transferring from digital CAD to an SL part, shrinkage happens the first time; transferring from pattern to tool, shrinkage happens the second time; transferring from tool to part, shrinkage is factored in a third time.

The CU casting process is done using either low pressure casting or vacuum casting; therefore, the formative properties of the part are not the same as injection molding in which molten plastic is injected at high pressures and temperatures producing compact, homogeneous, and better quality parts. Pressure casting will squeeze down the air bubbles in the material to a microscopic level in a pressurized chamber so that they have no effect on the part. Vacuum casting will prevent the air bubbles, as the material is poured in a vacuum chamber.

Complex geometries may require special handling and multiple pours to encapsulate all the geometry. Curing time varies with part size, geometry, and materials used, ranging from minutes to twenty-four hours per pour.

### Step by Step, Making an RTV Tool and a CU Part

Designers need to fully evaluate their design before making a pattern. Typically, if you try to save money by making your pattern during the design check, it causes problems. When design issues are overlooked they get duplicated in the part. All designs should be corrected first to get an error-free CU part.

This is the process for making an RTV tool and how CU parts are made.

## CAST URETHANE PROCESS

| Parting Line | Master Pattern | Sprue & Vents | RTV Silicone Casting Resin | Frame (BOX) |

1. Select the pattern.

To make a tool or mold, the engineer chooses a physical pattern based on strength, surface smoothness, and design objective. The tooling will be a negative or inverse replication of a basic pattern, typically an SL part. The engineer discusses possible issues with the molder, such as material shrink factor, tolerances, and special features. Keep in mind that a standard, unpolished SL will not make a good pattern. Pattern surfaces need to be glass-smooth to make the corresponding tool very smooth. A very time-consuming, high-quality finish is applied to the SL pattern by an expert craftsman.

2. Make the RTV tooling (silicone mold).

Next, the molder defines the pattern's best parting line. He sets up the pattern oriented to the parting line, then pours the rubbery RTV substance over the pattern and lets it cure. When the RTV substance has solidified, the molder cuts out the pattern around the parting line, then physically pulls the tool off the pattern.

3. Cast the part.

The molder sets the inverse-shaped tool on a level platform. He mixes together two reactive liquid materials to make ure-thane, which he then pours into the tool. The part can be cast with minimal to no pressure. If pressure casting is required, urethane material is poured into the tool and put in a pressure chamber. The pressure squeezes the air bubbles in the part to be as small as possible so they will have no affect on the part. If vacuum casting is required, urethane material is poured in the tool while in a vacuum chamber so that additional air is not introduced into the part during the pouring process. The vacuum sucks air out of the mixture to ensure solid formation within the part.

After solidification, the material sets in the tool anywhere from five minutes to several hours. This wait time allows the parts to develop enough strength to de-mold without breaking. Because parts are fragile at this stage, they must be handled with care. The chemical reactions are still occurring throughout the part at a molecular level. Next, the part is transferred to an oven for post-curing for several hours until full cure is achieved. The operator de-molds the newly cast part from the soft RTV tool and gets ready to make the next part.

## :: CU Applications—A Polymer Chain of Fools

Look around you: polymers are proliferating madly. The CU process touches nearly everything in your environment. From the orange plastic cones on the highway to the old-fashioned ice cube tray in your freezer, CU is used to make a wide variety of parts in every industry from jewelry settings to Space Station components. Better yet, take a wide-eyed walk through Wal-Mart or Target to see just how much of your world is made possible by polyurethane, the ubiquitous matter.

### *Benefits—There is No Buzz*

While CU is something to get excited about under the right circumstances, there is no industry buzz. The CU process has become a universal, standard method in manufacturing. We could get really excited and grateful about highways, electricity, and running water, but after awhile, we absorb these miracles into the background of our existence.

The main excitement of using this bread-and-butter process is that **CU offers you low-cost test samples prior to committing to more expensive tooling.** Design changes can continue while the rest of your collaborative team gets what they need early in the process. With early design CU samples in hand, your team can accomplish product validation, get colorful, authentic-looking trade show or marketing visual aids, try out form-fit-function with the complete assembly, and test fixtures for production processes. In the reverse engineering world, CU is a mainstay for preserving a master part in situations where you may have only a few parts left and need to keep a master or "go-by" ready. CU is a lifesaver for making patterns of parts that could be destroyed in the reverse engineering process.

## :: Boring but Necessary—Understanding CU Limitations

The CU process is a blend of art and science. As such, it requires a skilled, experienced craftsman to understand how your design tolerances will hold up in spite of somewhat unpredictable shrink rates and moody chemical reactions. Mother Nature is most definitely a player in the chemistry of making your parts. Here are a few pointers in case you plan to start making CU parts under a full moon.

### Part Features

In CU parts, detailed features are difficult to form due to the pouring process. They can easily turn out deformed or incomplete with voids. Difficult to cast, small holes must be added to the part after it is cured. This secondary process adds time, cost, and complexity to the part creation.

### Tolerances and Accuracy—Shrink Factor x 3

Don't expect CU tolerances to be as tight as production tooling. RTV tooling is considered a low-tolerance tooling method in which allowances are required. Because the CU process has difficulty managing tolerances, skillful craftsmen are needed to ensure the best outcome in a three-step process. Since this process requires the creation of a pattern, a tool, and a part, there are three unique materials and processes that have their own tolerances. The master pattern holds a shrink factor from the RP process; the RTV has shrinkage from its process; and each urethane material has shrink factors in the casting process. Additional shrinkage happens in urethane materials according to the design geometry; therefore, the outcome can be very difficult to predict. Material data sheets are misleading since they are based on a test sample only.

### Size Matters—Are We Surprised?

When only a few parts are required, **the CU process can be more economical than buying tooling for very large parts.** All sizes are available using CU; however, a tool can get to be big, heavy, and expensive.

## *Materials are a Nightmare—Polly Who?*

The word polyurethane typically refers to flexible foams used in car seats, sofas, bed mattresses, pillows, insulation, packaging, earplugs, resistant coatings, adhesives, sealants, and packaging. Mixed with the right chemical, urethane casting materials can also be formulated to imitate elastomers, high-impact materials, and glass-filled materials.

### • Material Types—Eggs or Ice Cubes

Several vendors supply many choices of materials for your CU part. Beware of material data sheets that omit more fact than they describe.

Because your CU parts will "behave" very differently based on your material choice, it is important to know the difference between thermoset and thermoplastic properties. Thermoset is like plastic but has more limitations. Using thermoset is like cooking an egg. Once cooked, it is done and cannot be "undone." Thermoset cannot be melted, and if heated too long, it burns rather than melts.

On the other hand, thermoplastic material can be melted and re-melted. Used to make injection-molded parts, thermoplastic materials are melted prior to being injecting into the tooling. Thermoplastic is like an ice cube that can be melted, refrozen, and re-melted.

Generally speaking, high heat is a challenge for CU parts.

## *How Will My Part Look?*

Many CU parts look as good as production parts, but quality really depends on the full orchestration of the three-step process, with multiple chemical reactions and curing times. Many uncontrollable variables announce themselves in the reactions among master pattern, RTV tool, and the part.

### • Surface and Post-Processing

Machined, sanded, glued, or painted, chameleon-like urethanes can take on the material and thermal properties, colors, and surface textures of the more expensive production parts. In other molding processes, texturing is done directly to the tool. However, texturing for CU is applied to the pattern first, then transferred to the tool in the duplication process.

Finish is any additional design detail that can be manually produced on the pattern. Typical plastic finishes are determined by tool texturing, a special process that uses acid etching. Only a skilled craftsman can copy the texture needed to make it look as close as possible to the standard texture. Other finishing touches include drilling holes and making inserts after the part is cast.

### *How Long Does it Take to Get My Part?*

Since there are three steps in the CU process, there are three internal schedules involved in making your part. Here's how it breaks out.

Making a smooth, high-quality pattern can take four to five days to produce and finish, unless you break the pattern and have to start over, which does happen on occasion. Also, if the customer is reviewing the pattern before moving forward with the design, then three to five days must be added for shipping and review time.

Making an RTV tool can take three to five days depending on the complexity of the part and the number of mold pieces required. Rushing the tooling phase is unwise. Because the laws of nature govern the outcome of the RTV chemical reactions, Mother Nature works at her own speed, no matter how much you need these parts to save your job and no matter how much money you throw at her to go faster. Better planning is the solution to avoid rush scenarios.

Once the tool is done, parts can be processed. Depending on the material used, making a part can take 10 minutes to 24 hours. Again, Mother Nature drives the chemical reactions according to her schedule.

If you add all three steps together, it is safe to guesstimate that your CU part can be made in 10 to 15 days.

The next manufacturing process, CNC machining, is another industry standard. **CNC is better for applications requiring high accuracy in metal or plastic.**

## ▓ CNC—The Intelligent Machine

### *QuickSMART*

**Definition:** CNC machining is a subtractive fabrication process that begins with a whole block of material and removes or "sculpts away" the unnecessary material, leaving only the desired part. Through the use of software and controls, the CNC operates automatically and efficiently and does not require human interaction during operation. CNC machines include other machining tools such as lathes, multi-axis spindles, milling machines, laser cutting, water jet cutting, and wire electrical discharge machines (EDM). The functions formerly performed by human operators are now performed by a computer-control module.

**Why You Need It:** To get the very best process for metal parts; to hold high tolerances; to get accuracy, repeatability, and reliability; to ensure high-quality output; and to get more material reliability and stability than chemical processes.

**Ideal Uses:** Low-cost method for cutting and shaping precision products, such as automobile parts, machine parts, and compressors; excellent for parts requiring high-accuracy applications in aerospace, industrial, and machining environments; process is mostly free of human errors and variations on accuracy and interruption;

low tooling costs required, typically jigs and fixtures; virtually any material (metal, plastic, wood, or foam) can be machined to make a part; and it can be used for trade show models, precision parts and tools, and one-of-a-kind, artistic designs.

## Subtract Your Way to Success

The genius Renaissance sculptor Michelangelo said that every block of stone contains a statue trapped inside it and the sculptor's task is to set it free. "I saw the angel in the marble," the great artist wrote, "and I carved until I set him free." This is what a CNC machine does with the help of computer-aided manufacturing (CAM) software. It subtracts material from a raw block of wood, metal, plastic, or foam until it "sets the angel free."

The introduction of the CNC machine, developed in the late 1940s, has radically changed the manufacturing industry. Keeping pace with advances in CAD and CAM technologies, today's CNC machines are controlled directly from files created by CAM software packages so that a part or assembly can go directly from design to manufacturing without a paper drawing of the manufactured component. Like HAL, the talking computer in the motion picture, *2001: A Space Odyssey*, a CNC machine can even be programmed to call your cell phone if it breaks down, detects an error, or just wants to scare the crap out of you.

### ▒ CNC Process

CNC machines cut away material from a solid block or "work piece" of metal, or plastic to form a finished part. CNC machines are typically used to produce large quantities of one part, although they may produce low-volume batches or one-of-a-kind items. A service provider's programming skill and knowledge of metal properties result in machined parts that meet precise specifications.

In the most basic terms, the CNC machine is a table with an automated three-axis mill that can drill in three directions, along an x, y, or z coordinate. The most basic motion for a controller is to move the machine tool along a straight line from point A to point B. CAM software is used to program the cutting paths of the tools to remove the material.

Manual milling was done for centuries, prior to the CNC, a class of industrial robot. Next came early numerically controlled (NC) machines that could perform simple calculations and moved in a straight line. Today's CNC machines can cut curvy organic shapes, thanks to fairly recent advances in CAD and CAM technologies. Machining curves in metal, plastic, foam, or wood is now as risk-free to the designer as cutting straight lines.

The term "computer numerically controlled machining" means that computer software and controls operate the machine efficiently by telling it which drills and cutters to use as it removes excess material without human interaction during operation. With raw material work pieces set up, it's not uncommon for a CNC machine to run parts by itself over a weekend, without operator intervention.

**CNC is a very reliable process that delivers a high degree of quality for accuracy and repeatability.** Tight tolerances are guaranteed. CNC is the most accurate machining process and is most independent of human error as it relates to accuracy and speed. Successful use of CNC depends on the engineer and machinist's ability to plan the part's creation process and generate programs to execute the plan. Low tooling costs are typically required for jigs and fixtures.

CNC has very few disadvantages. As huge, high-tech equipment goes, CNC does require a skilled planner and operator to process a part. It offers no economies of scale in the process, meaning it can machine only one part at a time. Repeatability between parts is accurate, but each part is unique and independent of other parts. Errors can occur when the part is almost complete, causing the whole part to have to be scrapped. Some geometries, like internal cavities, cannot be made by CNC.

### Step by Step, Machining a Part with CNC

Prior to machining, the engineer needs to communicate with the service provider to plan how the part will be built, how many setups the part requires, and to assess what jigs and fixtures will be required in the setup. The engineer needs to check on material availability and acquire it before machining starts. The engineer and machinist will review the CAD data together to prepare for any special challenges or instructions. Once your CAD file hits the machine shop, here's what happens, step by step.

First, the CNC programmer carefully plans and prepares the operation by reviewing your 3D CAD model or blueprints. Next, he calculates where to bore into the solid work piece, how fast to feed the material into the machine, and how much material to remove. He selects the appropriately sized drills and cutters for the job and plans the sequence of cutting and finishing operations.

Next, the CNC programmer turns the planned machining operations into a set of instructions, or "g-code" from a CAM program containing a set of commands for the machine to follow. These commands are issued as a series of numerical codes that describe where cuts should occur, what type of cuts should be used, and the speed of the cuts. The operator physically sets up the block of raw material, held in place by jigs and fixtures, on the CNC table. He then checks the programs to ensure that the machinery will function properly.

After the programming is completed, the CNC operators transfer the commands from the server to the CNC control module. Many advanced control modules are conversational, meaning that they ask the operator a series of questions about the task. CNC operators position the raw piece on the CNC machine tool and set the controls.

The machinist activates the program to select the right tool for the first portion of machining. The CNC mills away a section of material until complete. The machinist then changes the setup to mill away another section on the part. This is repeated until the part is finished.

## ▉▉ CNC Application—Cold Blue Steel

Over the last 15 years, as the evolution of plastics continues to grow and the ability to make tooling becomes faster and cheaper, the most significant transition in product development has been the growing movement to redesign metal parts as plastic parts. Companies that were designing and machining in metal are now tasked with redesigning parts in plastic to boost their bottom line. Designers now strive to combine the functions of several, simple metal parts into one, complex plastic part. CNC is better for simple and prismatic (blocky) parts, while plastic parts are better suited for complex, organic shapes. Also, plastic parts can incorporate many features, while CNC parts typically require more unique parts to handle the needs of the product.

In the US, the number of CNC programmers has dropped, but reduction in manpower does not mean a decline in productivity. Thanks to advances in CAD and CAM software, one programmer can now do the work of two or three programmers.

Looking at the global picture, the US will not be outsourcing CNC services to China. While there are literally hundreds of thousands of CNC shops competing in South China, it's an impractical resource due to the cost of shipping raw materials and the finished product. If it were practical to take CNC work offshore, companies would be doing it.

### Benefits—The Literal Buzz

CNC operators are good listeners. They can interpret the literal buzz of a high-speed cutting tool to adjust cutting speeds and reduce error. They are trained to tell good vibes from bad to ensure part quality.

There are many benefits associated with this high-tolerance machine, used primarily by industrial clients in aerospace, military, and heavy-duty equipment.

The best applications for CNC include prismatic metal parts or plastic parts requiring very high tolerances. Cost wise, CNC can yield a much greater efficiency on blocky, voluminous parts than other processes. CNC is ideal for the customer who needs a part made from a special material, like thermoplastic, but does not want to invest in the tool to produce a plastic part. It is also practical for very low-volume quantities or when the end-use production process of choice is CNC.

## ▓ Boring but Necessary—Understanding CNC Limitations

Any way you look at it, CNC machining is a wonder. A largely automated process, it offers high accuracy, fine features, solid speed, and material friendliness. This is one situation when saying "it bores easily" is a good thing.

### Part Features

CNC offers great flexibility on part features from very fine detail up to coarse drilling. However, CNC cannot handle complex undercuts or internal cavities that cannot be reached by a drill bit.

### Tolerances and Accuracy—Real World

CNC machining offers the most accurate tolerances of any manufacturing process—typically ± 0.002 inch (two thousandths).

### Size Doesn't Matter

The CNC has no real limits on the size of a part. Large CNC tables are available to accommodate large parts or pieced-together parts. Bigger parts require more special equipment, which adds to cost of material and machine time.

### How Long Does it Take to Get My Part?

CNC requires a great deal of planning and processing. Typically, it can take a week to get the first batch of parts processed and machined. After setup, subsequent parts take only the machine time. Very simple parts can be made in two to three days, but typical parts can require seven or more days.

## Data Requirements

CAD models are required for CNC machining. Typically STEP or IGES files are the output of the CAD system that feeds into the CAM software.

### A Very Tall Tale—Johnny Quickparts Orders Meat and Potatoes

Mr. Overrun came dashing in to Johnny's dank cubicle with a new crisis. An aerospace client needed a high-precision, super-thick, stainless steel bracket. He would use it to test the "Three Little Fs"—form, fit and function—of his latest "smart bomb." At the same time, the aerospace client's marketing dummies needed 10 sample brackets to sell advanced orders to foreign militaries. The client emphasized a rush on his chunky bomb bracket. With world tensions escalating, his shareholders wanted to "make haste" before world peace caught on.

Despite waning biorhythms, Johnny told Mr. Overrun he would jump right on it. He'd just gotten back from a month of travel throughout the US, during which time he was held captive by a fussy client at the Chateau Nouveau, a resort known for ice sculpture hotrods and over-the-top nouvelle cuisine. All the high-tech recipes and layered desserts had been confusing to Johnny's soul. He was glad to get back home. Secretly, he planned to leave work early, microwave a frozen TV dinner, and watch '50s reruns in his flannel jammies.

Spinning the customer's CAD model around in the 3D space of his workstation, Johnny recalled his terrible faux pas the last night at Chateau Nouveau. He had insisted on meat and potatoes, nowhere on the menu. After a 30-minute tableside inquiry, the insulted chef deduced that Johnny was just hungry. The chef prescribed a roast beef and potatoes dish designed to leave a velvety "mouthfeel" on Johnny's palate. "Mouthfeel," Johnny learned, could be achieved only with large amounts of animal fat, to which he said, "Bring it on!" He had thanked the rattled chef profusely for, just this once, aiming low. There was a time and a place to forget the frou-frou and go with basics that had stood the test of time.

Johnny snapped out of his traumatic travel memory and gave a mind-meld to the customer's smart bomb bracket. This time, *haute technology* was not needed. Going back to the basics for this project was best. CNC machining would provide a super high-precision, stainless steel bracket for his customer's engineering needs, and the CU process would provide 10 low-tech, inexpensive brackets for the marketing dummies. He emailed Helen Helpalot and asked her not to call him after his upload because he had "laryngitis." She replied with a sunny email that said, "Your order will be ready in 10 days, and by then you'll feel like talking."

Johnny smiled at this most perfect woman he had never met. He uploaded his file, emailed his boss a lame excuse for leaving early, and with a flick of a switch, shut down his computer. Even Johnny Quickparts took a wellness day every now and then.

## ▓ Saving Time, Saving Money—Saving Thousands with Alternate Processes

### *How Do I Save Money Using CU?*

Before purchasing CU services, study the price per part offered by low-volume tooling as an option. Find out the quantity at which it makes more sense to go with CU. Make sure that you understand the functionality of the parts before entering into a process that will not work—for example, in high-temperature environments. Always buy your parts from a high-quality provider so that you don't have to risk re-doing the project. Since strong artistic skill is needed in this process, you want experience working for you. Always use instant quoting to get your best price.

### How Do I Waste Money in the CU Environment?

Learn material limitations. You will waste money if you expect thermoset parts to behave the same as thermoplastics parts. Using a bad pattern, such as an outdated SL model to cast the tooling, will also create waste. Don't order too many parts at one time. Only buy a few parts from a tool, as it will most likely go through design changes. Don't expect the tool to last for long periods of times after it is made. An RTV tool does not "rest" well on the shelf, and silicone continues to dry out and degrade over time.

### How Do I Save Time Using CU?

As with all of the fabrication processes, understand real-world tolerances, know your materials, and double-check your CAD revision date before submitting it. Research the limitations of each process, especially regarding part features. With CU, small holes should be drilled in after casting is finished. While this is fairly easy to do, it does add more time to the job.

Always review the master pattern, even if you are guessing that it's fine. If you don't catch the error up front—be it a tolerance, feature, or design issue—the error gets duplicated all the way through the three-step CU process. There are risks to reviewing the pattern, such as breakage during review or shipping. Patterns can also warp quickly due to heat or moisture during the process. Always use instant quoting to save on valuable engineering man-hours, and always handle patterns with care.

### How Do I Waste Time in the CU Environment?

Typical time eaters result from having the wrong expectations. CU parts will not behave like thermoplastic parts in a functional environment. If you wait the two to three weeks for your parts and they fail, you have to start over. Poor assessment of mating part tolerances will also cause rework, as will poor communication with your provider about your requirements for the functionality of the part. Poor planning or no planning usually turn into an emergency, so planning is always a good thing.

### Savings with CNC

Be sure to use CNC only when you need it instead of using another process that may be much slower. Proper planning time is essential for good results with the CNC process which requires a balance of skill, art, and technical process. Remember that the CNC machine is controlled by a computer that follows your instructions, good or bad. Avoid changing the design after the part has started the CNC process. Be sure to buy only the number of parts necessary. Finally, CNC is a natural money saver. Once programming is done on the first part of a batch, it doesn't have to be repeated.

### ▓ The Keys to a Vintage Studebaker

By now you know that sometimes a high-tech solution is overkill for the application. It's ok to go back to the basics. You've been introduced to two manufacturing classics that have stood the test of time. The CU and CNC processes are admired as the enduring workhorses of the industry.

You've learned that CU is a three-step process, and you know why you always check your master pattern before you make the tool. You know that the CNC machine is a class of robot that provides super-high tolerances and can cut steel. You know that thermoset and thermoplastic properties are as different as eggs and ice cubes. You've watched as Johnny deals with a bad hair day and solves all of his engineering problems by going back to the basics. Here are the keys to some classic manufacturing wisdom.

As Johnny would say,

**"It should work."**

# Chapter *6*
# Low-Volume Injection Molding

## TO BRIDGE OR NOT TO BRIDGE

*"As with most things in life, folks tend to focus
on the end game, the score, the finale, but
choose to ignore the many critical steps and
decisions that are made during the journey."*

### *QuickSMART*

**D**efinition: Low-Volume Injection Molding (LVIM) is a manufacturing method that creates injection molds or tools to produce functional parts from thermoplastic in short runs of up to typically 50,000 parts. Significantly faster and cheaper, LVIM offers the same quality, accuracy, and tolerance as production tooling, but without 2D drawings.

**Why You Need It:** To reduce wait time; to compress production time; to make parts while your production tool is being produced; and to deliver parts to your customer in two to four weeks, instead of eight to twelve weeks with a standard production tool.

**Ideal Uses:** Simple, single-cavity tools; a "bridge tool" in aggressive product development schedules; low-volume requirements for applications with a short lifespan; and is sometimes used to test heat-resistance and functionality in end-use materials.

## Low-Volume Injection Molding Basics

Throughout this book, you've learned about options for making a prototype or part using Stereolithography (SL), Selective Laser Sintering (SLS), Fused Deposition Modeling (FDM), Cast Urethane (CU), and Computer Numerically Controlled (CNC) machining. These fabrication processes are all used to verify that a design represents the intent of a product. Now it's time to get tooling made. From tooling, you will produce end-use plastic parts. Your choices are LVIM, production tooling, or a combination of both.

LVIM has existed as long as production tooling, well over a hundred years. In 1868, John Wesley Hyatt, a US inventor with hundreds of patents, was the first to inject hot celluloid material into a mold to produce billiard balls. He was looking for an alternative material to traditional ivory. The injection-molding process remained the same until 1946 when the first screw injection molding machine revolutionized the plastics industry. Today, almost all molding machines use screw injection molding to heat and inject plastic into tools or molds.

> **QuickTip:** The labels "tool," "tooling," "mold," "mould," "molding," and "moulding" are all used interchangeably throughout the industry, causing great consternation to outsiders. Similarly, a "tool maker," a "mold maker," and a "mould maker" all make the tool. Additionally, a "molder," a "processor," and an "injection molder," make the parts. It's all good!

## Contrasting Low-Volume to Production Tooling

The term "Low-Volume Injection Molding" means different things to different people. To a designer, it may be a tool that is used to make a relatively small number of parts. To a tool maker, it may be a tool that has been built to demonstrate a strategy for making the production tool for a complicated part and to verify if it will perform as anticipated. In the current world of product development, the use of LVIM is a critical strategy to expedite development. By leveraging the advantages of injection molding, the developer is able to get his or her product to market faster without jeopardizing the result or increasing risk of failure in the complicated world of production tooling.

Knowing the key differentiators between LVIM and production tooling will help you make decision tradeoffs as your product moves forward into the most critical and expensive phase of manufacturing. Being lower in cost and much faster to produce, LVIM means that you can get your product to market faster. The simplicity of the LVIM—usually a single-cavity design—allows faster creation, whereas a production tool with many cavities takes much longer to build. Low-Volume Injection Molds have a short life span and can withstand making up to 50,000 parts, while production tools live a long time and have the strength and durability to make millions of parts. In low-volume production, the design goal is to keep the tool simple and use manpower to help process the parts since the volumes are lower. In production, the tool is designed to be mostly automatic which reduces the cost per part. LVIM typically does not need water lines; however, production tooling does require water lines for cooling and would be included if it were to validate a tool design. Water lines add complexity, time, and expense to the production tooling process.

Additionally, LVIM typically does not have as many moving parts, actions, or features as a more complex production tool. Lastly, LVIM is typically made of aluminum or soft steel, requiring two to four weeks to make, while production tools are typically made of high-quality steel, and are deliverable in eight or more

> **QuickTip:** Injection molding is the most common manufacturing method for making plastic parts. A tool maker creates the tool from steel or aluminum. Under high pressure, molten plastic is injected into the metal "tool" or mold cavity, filling the inverse or negative space to make a positive-shaped part.

weeks. This key difference in delivery times makes LVIM a priceless "bridge tool" technology, enabling part delivery much earlier than the production tool.

### Impact of CAD and CAM Advances on Product Development

Limited in the past 20 years, the product developer's option to use the LVIM process opened up as a result of the growing rapport and eventual marriage between CAD and CAM technologies. Evolving CAD solid models merged with changing CAM technologies that were enhanced to handle these complicated models. Through the ever-deepening marriage of CAD and CAM, the product development and manufacturing worlds have now absorbed the reality that a product can be designed and produced in a matter of days. This time compression is a direct result of electronically contained data in a file that is now transportable to all phases of tool making. Significantly valuable to the product developer, this technological evolution is nothing short of amazing.

In the early '90s, a product developer's only option to fully produce a part—using the end-use process with the end-use material—would be to buy the production tool and hope the design worked. At that time, it took 12 or more weeks to have production tooling produced at a high cost. As an example, imagine that it is 1990 and a product developer needs a new widget made out of a special thermoplastic to test the design. He has models made from wood or even machined plastics, but these prototypes do not represent the final part very well. Suddenly, a new process called Stereolithography appears and promises that you can now get your "plastic" part just as you designed in a few days for a fraction of the tooling costs. At this point, the product development world responded with a big "Wow!" to the rapid prototyping (RP) revolution. While time and cost impediments had spawned the need for prototypes to verify designs, the advent of reduced dependency on production tooling to test the form, fit, and function of a part. This change forced the tooling industry to regroup and evaluate how to stay competitive. Eliminating production tooling time and expense, LVIM became an alternative solution, and resulted in a major technological advantage to the manufacturing sector.

## • The Wall Tumbles Down

Prior to the marriage of CAD and CAM, a virtual "Berlin Wall" divided the disciplines of engineering and manufacturing. Engineers designed tools in 2D and sent drawings over "the wall" to be interpreted by manufacturing in its own language. When CAD and CAM merged, the wall came down and reorganized the process flow while eliminating wasteful steps. A paradigm shift ensued, proving that tooling could be made in days instead of weeks.

Since that time, product developers have had over 15 years of development to assess the evolutionary promises made by technology to discover which ones were kept, and more importantly, how, why, and in what context. By now, the limitations of each process are fully known.

As necessity and competition drive all things to be better, it turns out that injection molding is also competing with additive processes that have displaced many molding opportunities. While tooling did not change much in the 60 to 80 years prior to this phase, the interrelationship of CAD and CAM now provides clear technological advantages while forcing the old tooling mindset to upgrade at warp speed.

> **QuickTip:** Never shake a baby and never weld a tool before texturing.

Amazing to the younger generation, you can still find "dinosaur" tool makers with their heads in the sand. Outmoded tool shops from the '50s and '60s, once big fish in a small pond, don't realize the Ice Age has come and gone. The old guard mold makers will actually argue that none of this "new technology" works or even exists, which is sort of like arguing about whether there is an "information superhighway." "We don't need nunna that," is their typical refrain. Of course, new manufacturing processes are available and they do work. Widely accepted around the world, LVIM is used every day by product development companies. Clearly, only those who embrace the new paradigm manufacturing will be victorious in the business of the future.

## *More about Production Tooling*

"Production" is a relative and nondescript word that means different things to different people, depending who they are and what they need. By now you know how to make and verify individual prototypes and parts as quickly and economically as possible. However, **the purpose of product development is to produce an entire product,** usually consisting of an assembly of individual parts. At this stage of tooling, a product developer will have to make tradeoffs between LVIM and production tooling, or both. Since production tooling is the final, most critical, and most expensive step of manufacturing, a working knowledge of tooling options is essential for choosing the best production path for your product. A product developer will invest thousands of dollars on production tooling so he can make thousands of parts for a product. All of the previous costs in the design and test phase will be only a fraction of the total product development process.

> **QuickTip:** Production tooling is the tooling or mold required to make injection-molded plastic parts. The plastic parts are the production parts required to assemble the end product for the consumer.

What makes a part "production" versus "non-production" is a judgment call, usually implying quality standards. With production parts, the highest levels of quality and functionality of the part become critical. Production also implies higher quantities of parts.

Both LVIM and the more complicated production tooling are made by essentially the same process as outlined below.

## ⠿ Process of Mold Making

Making an LVIM is a fascinating process in which you create something to create something else. One of the major challenges in the process is that you must create a mold or tool that can be used as a receptacle for molten thermoplastic that holds the inverse or negative shape of the part you desire. While this sounds simple enough, some special knowledge is required.

Making the physical tool is just a piece of the battle. The part geometry you design must be conducive to the molding process, and the end-use material must be conducive to the part as well as the mold. The many variables of the process—design, materials, actions, and expectations—make the process of getting from tooling to parts a challenge.

The more efficient LVIM process is similar to the tool making process in that it has existed for over a hundred years. As with sculpting, the tool maker eliminates what is not required and keeps only what is essential.

**LOW-VOLUME INJECTION MOLDING (LVIM)**

Core Mold · Water Lines · Sprue · Cavity Mold · Ejector Pins · Vents

### Step by Step: Making an LVIM

Mold making is a complex science that requires a high level of expertise in design, materials, and physics, along with artistic and intuitive insight, all part of the mold maker's trade. A highly valued and specialized craftsman "begins at the beginning." He starts with a great plan for a well-designed part and follows through with flawless execution, resulting in a very smooth, high-quality injection mold.

If those of you who are engineers are now scratching your heads, you are not alone. For some reason, this valuable mold-making module is not taught in engineering school. Here's an important addition to every designer's knowledge base. Step by step, this is how you make a mold.

1. Plan how to make the mold.

   a. Assess the part for the injection-molding process or Design for Manufacturability (DFM)

When designing a mold, make sure it is conducive to injection molding. The design process for plastic parts is critical, taking into account the "moldability" of a shape. With today's easy-to-use CAD software in the hands of very "green" designers, it is common for parts to be designed that can be prototyped successfully with SL, SLS, and FDM, and accepted by the customer, yet still unable to be injection molded. This costs your company thousands of dollars in errors, issues, and lost opportunities. Early in the process, the expert tool maker closely considers all that could go wrong with a design. Defects that result from poor design and require costly rework arise as lack of draft, parting line problems, poorly fitting ejector pins, poor materials selection, feature deformation, and tolerance errors. The next steps happen electronically in CAD during your design process.

   b. Determine the parting line of the part.

The tool designer visualizes where the tool will come apart in two halves for part release. The line that is formed at these mating surfaces of the tool is called the parting line. The parting line choice is important because it affects the aesthetics and possibly the functionality of a part. Also, the parting line is subject to variability as the tool is processed. As the tool wears, a lot of activity occurs at the parting line where the halves meet. The parting line is susceptible to issues of deformation that occur when the mold is not precisely mated to close completely or is pushed apart under pressure. The resulting gaps fill with unwanted material called "flash."

c. Create the part negative from the mold halves.

Working in CAD, the mold maker orients the part for the parting line within a virtual block of material. Next, he does an electronic subtraction of the part to leave the negative shape of the part in the work piece. This will provide two new parts, core and cavity, that contain the negative or reverse portions of the part being designed. The core and cavity meet at the parting line. This process happens simultaneously in CAD so that it appears as a single piece.

d. Determine sufficient venting for the mold.

The tool maker visualizes and designs the best escape routes to vent air from the tool as it is filled with molten plastic. Vents are needed to prevent voids and bubbles caused by trapped air. When the injection mold process begins, heated plastic quickly displaces air from the tool. The vent allows the air to escape under pressure. The venting of a part is typically tuned during the mold testing which may require new vents or changing the vent design.

e. Determine the best ejection system for the mold.

After the plastic is injected into the mold, the part remains "stuck" until the mold halves are released or pulled apart. The new plastic part still needs help to loosen and eject from the mold in the same way a cake needs help in releasing from a cake pan. In injection molding, the process uses ejectors to push the part out of the mold. Ejectors are strategically located pins that push the part from the mold after it has solidified and is very hot, without deforming it. Ejection must be designed to be a part of the process without human intervention. If the ejector system is not done well, the part will stick in the mold and possibly cause the part to deform with extraction.

2. Machine the mold halves with CNC and EDM.

After the tool is electronically designed and all key decisions have been made, the machinist's physical work begins. The CAM software technician processes the data for the mold halves to be machined with CNC, making this process very easy and versatile. Also, some features are processed with an Electrical Discharge Machine (EDM) using an electrical charge to burn away the excess, unwanted material. Today's software is highly sophisticated and easy to use. Built on the same interrelated model as the CAD data, the CAM output will change automatically if the CAD data changes. High-speed CNC machines today can also cut metals faster, but the time advantage is really just incidental. The real power is in the CAM software and the CNC process.

3. Mate the halves for fit.

After the mold halves have been completely processed and machined, the tool maker mates them together. Mating surfaces is a high-precision process. The end result must be very close to perfect, with no gaps or misalignments. There are many tricks of the trade, such as an ink stamping process called "bluing." Bluing is used to check for the transfer of ink to the other half of the mold to ensure full mating of mold halves. (An interesting side note is that the US typically uses blue ink while China typically uses red ink.) The critical need is for the surfaces to mate perfectly before continuing the process. If not, only expensive future rework can fix this error. Mating has a major impact on the overall quality of the parts that come from the mold, and it can add extra "features" from the mismatch called "witness" lines. If a small gap between the two halves goes undetected, then extra material will be squeezed into this gap, leaving obvious traces that may ruin the part. It is common that during the mating of the mold halves that the molds need to be polished so they are very smooth to produce the best parts. Polishing is a very time consuming process.

4. Assemble the mold.

Mold assembly is where a real time drain can occur. After the mold halves have been completely mated, they are assembled with supporting hardware much like a 3D puzzle. Supporting hardware includes fitting every piece that is required to make the mold workable, such as ejector pins, actions, and alignment guides. By assembling the two halves with all hardware, the tool maker ensures that, for the first time, all pieces are available and assembled correctly. Time drain can occur if parts of the mold have been forgotten or were incorrectly made, such as slides or lifters being too big or fitting too loose. A small but critical error like this stops all progress while the seemingly insignificant pieces are reworked.

5. Install and test the new mold.

The trial run with the injection-molding press is where the "rubber meets the road." This step reveals whether all of your previous work comes together or falls apart. The molding processor takes over from the tool maker and hangs the mold in the press. He shoots hot plastic into the mold as a trial run to see how it performs. He hopes that a perfect replication of the part design will result, but this would be uncommon on the first trial. As with most creative processes, iterative changes are required. The first shots are used to identify tooling problems or design issues. A plastic part stuck in the mold can mean many things, most likely that the ejection system is not working or there is not enough draft designed in. At this point the tool maker's "artistry" is required to diagnose the issues and make the required improvements to the mold. It is a challenge to predict how long it will take to get the mold just right.

6. Make parts from the mold.

Once validated, the mold is now production-ready. The operator installs the mold in the injection molding press. Plastic pellets are funneled from a hopper, then heated and forced under extreme pressure into the mold cavity. Within seconds, the injected plastic solidifies into the shape of the part. The mold then opens automatically and ejects the newly formed part. After the mold ejects the part, the process repeats.

## ⠿ LVIM Applications—Everything is Everything

Look around you. Practically everything is an injection-molded part. If you tear apart any of your handheld gadgets—cell phone, tape recorder, computer mouse, electric toothbrush—you will discover a multitude of injection-molded thermoplastic parts. From the handle on your lawn mower to the produce drawer of your refrigerator to the buttons on your radio, you sit at the center of a plastic injection-molded universe.

### *Industry Overview—A Nerd's Eye View of LVIM*

With current technologies and the growing acceptance of LVIM, applications of this process continue to expand. LVIM has now become a standard element of the product development process. Decades before LVIM, a production tool was predominantly focused on proving that a part could be molded successfully. In other words, product developers had to use full-on production tooling to validate a part; there was no intermediary refinement process to see how the part would "behave" in reality. But the LVIM process has evolved significantly with the use of CAD and CAM technologies. It is now considered a very useful technology in the iterative development process.

There are many ways of using this process to get your products to market faster. Product developers often use a "bridge tooling" strategy that includes both LVIM and production tooling, either in parallel or in sequence, to support their goals.

Applications for LVIM are found in every industry sector: industrial, automotive, medical, lawn and garden, and consumer electronics. Close tolerances and high-end appearance are ideal for today's short run projects.

## *Benefits—The Buzz is Real*

In defining your best strategy for compressing product timelines, consider the benefits of LVIM. Product developers are catching on to its powerful bridging capacity used to bolster the front end of larger production projects. **LVIM provides the only solution for creating a few real parts for functionality testing in the end-use material.** This short-run tool is priceless when it comes to garnering investors in early market evaluations. Using LVIM as insurance provides additional safety to your bottom line.

### • Short Run Needs

Product developers use LVIM for short run needs when they need a few thousand parts to get the product to market. Since the LVIM process is fast and cost-effective, it's a great way to get low volumes of parts in the end-use material and beat your competitor to market. LVIM is useful in many situations in which you may not be sure of the market's demand for your product. It's also useful if you're still trying to overcome design or technical challenges. Essentially, an injection mold is a dispensable or disposable tool that has the sole purpose of creating a few parts that look like the production parts.

LVIM is commonly used for short runs in medical and industrial sectors, situations in which the product already has a very low-volume requirement and may have many phases of iterations planned into the design. These application types require much process flexibility and the ability to get parts fast and economically. Short run applications do not require a production level tool to meet the needs of the product.

### • Real Functionality Testing

Product developers use LVIM to get a real, functional test unit to verify the product by getting parts made from the specified thermoplastic material. Rapid prototyping processes would not work because they do not produce parts in end-use materials.

LVIM represents an amazing advance in the way products are developed. Not only does it allow the prototyping of parts in the actual end-use material, but also the parts are made in a similar process as the final production parts. Therefore, you are getting an excellent test of how your actual parts will look off the production line.

Using LVIM is a very common requirement in areas in which the part will be used in harsh environments, high temperatures, or at high loads. Product development teams need to know exactly how that part will react, but they don't want to invest the time or money required for a production tool that will need to be replaced. This process allows them to gather new information in prototype testing before investing in production tooling. While the production tooling approach is a very expensive way to develop a product, in some situations it is unavoidable as there is no other way to actually produce the product and assimilate injection molding without the final process. With the use of LVIM in today's world, the cost is very reasonable and the time significantly reduced.

### • Bridge to Production

Product developers use LVIM as a "bridge tool" or transition to get some of the product to market while their production tooling is being made. LVIM can be a very powerful way to augment a product development strategy. As engineers know, many unknowns and potential risks to the schedule can occur when the production tooling process gets started. Bridge tooling makes it possible to mitigate the risk associated with the production tooling schedule when an LVIM is made in parallel with the production tool. The product development company can have the latest version of the product started with an LVIM, which typically takes two to four weeks to process. At the same time, the mold maker starts the production tooling process for the same part. By the time the LVIM is done, the design for the production tooling is well on its way—the CNC is ready to start cutting a block of aluminum into the core and cavity of the production tool. The product development company can then begin assembling parts and shipping products to market while their production tooling is being produced—all without risking any schedules or forcing the production schedule to be more aggressive than it needs to be.

### • Insurance

Product developers also use LVIM as a backup when they are not certain of production tooling schedules. Often used as "insurance," LVIM is needed when the product developer is developing a complicated part, using an exotic material, or trying a new supplier. By leveraging the economics and speed of the LVIM process, the product development company can feel assured that it is fully leveraging the resources available without risking the future of the product. Like insurance, LVIM covers the risk associated with the challenges of the product.

### • Market Evaluation

Product developers use LVIM to assess the product prior to investor commitments. The use of LVIM is an excellent way to evaluate the market for a product. It is not uncommon for a product to go to market and be ergonomically unacceptable to the consumer if features are inaccessible to the user. The LVIM process allows product development companies to get real-world data on their product before investing heavily in a production tool that can produce millions of parts. With this valuable marketing feedback, companies can fine-tune their design to perfection so that when they do go to market, product success is guaranteed.

### A Very Tall Tale—Johnny Quickparts Goes to China in the Year of the Pig

On a frosty December day, Mr. Overrun seemed overjoyed about getting an order for 500,000 purple plastic Easter Bunny Radio (EBR) housings. The housings had to be engineered with a tolerance of ± 0.005 inch (five thousandths) and the housing halves had to snap fit tightly, able to withstand being thrown by a mad toddler. With more fanfare than usual, Overrun asked Johnny to lead this important project using China. Acme's president, the rarely seen golfer type, had just pinned Overrun in the men's room and asked him why he hadn't sourced in China yet. Overrun excitedly told Johnny that this new mandate came from the president's golfing buddies, so it had to be true. Acme could save millions in China! "Hell, it can't be that different from the US," Overrun said. "Enjoy the junket and bring back a few numbers to satisfy his imagination." Johnny agreed with a handshake. The boondoggle was on.

The customer was none other than the world's most beloved discount retailer Pal-Mart, selling everything from apples to zippers. "Project EBR" had to be assembled, packaged, and shipped, no later than March 1st. The bunny radios would be on the shelves eye-to-eye with smart shoppers well before Easter. To reduce production risk, Pal-Mart ordered two lots of 500,000 from two competing product developers, Acme Design and its dreaded competitor, SCROO-U Unlimited.

Word got around at the American Society of Mechanical Engineers (ASME) chapter meeting that SCROO-U's lead engineer was all puffed up about beating Acme to China. Johnny's nemesis, Todd Hubris, was really sharp and good-looking to the point of being scary, but he had no soul. Johnny considered him an arrogant know-it-all and had always politely refused Todd's sarcastic professional invitations to Inventors Pizza Night.

The next morning Todd left a snide but chummy voice mail for Johnny, implying that as engineers they were both above competing against each other to please their company management. Then Todd spilled the beans about his plan to quickly get SCROO-U's bunny radio mold made in China and produce parts there as well: "Eight weeks for tooling, T1 (trial 1) by Jan 30, shoot parts by Feb 5; deliver parts by Feb 28...then to the Bahamas with my new girlfriend. Did I tell you her IQ is 188?" Todd was rude, rigid, and rough—the three R's of any brewing disaster. As a rule, Johnny did not gab with "professional poison" and did not return hyped voice mails meant only to help make Todd feel *big*. Besides, Todd was dumb enough to give away his entire plan for probable disaster with unproven, unknown, uncertified factories in China. Todd's blabber-mouthing inspired Johnny to win this war quietly and confidently. The battle to win Pal-Mart forever had officially begun, but Todd didn't even smell the blood!

Coolly dissecting the enemy's puffed-up production plan, Johnny could see that Mr. Hubris held a number of incorrect assumptions about manufacturing in China. He had indeed put all of his Easter eggs in one basket. His decision-making left SCROO-U quite vulnerable to utter failure in far-flung regions of China. Worst of all, he had sounded way too sure of himself, to the point where Johnny was having flashbacks to his diaper days. Granny Quickparts used to preach about how "Pride goeth before a fall." In efforts to confound the enemy, Johnny wrote Todd a quick, "professional pal" email which said:

*Hey Hubris,*

*When in China, never insult, put down, embarrass, shame, yell at, or otherwise demean a person. If you do, they will lose "face." In ancient times, a Chinese warrior chief, after losing a battle, would commit suicide if he lost "face."*

*Get Wise,*
*Johnny*

Johnny put on his headset and downloaded snappy Easter tunes to inspire a winning solution. A "bridge tool" strategy would require making two tools: an LVIM for short-term success and a regular production tool to meet the 12-week schedule. This bridging strategy would doubly ensure the success of "Project EBR." He could relax on his first international junket knowing the job was foolproof in spite of any mishap in China. He also hoped his global savoir-faire would impress Helen Helpalot of Quickparts.

Johnny checked Pal-Mart's CAD file for design for manufacturing (DFM), then uploaded it to Quickparts.com. Within 10 seconds, Helen called him squealing in delight. She praised his high-level global thinking for ensuring early delivery while cutting costs. A double-barrel production approach would feature the first tool, an LVIM made of aluminum, as the insurance factor. The LVIM, made in the US, would be ready in four weeks, and Johnny could deliver 100,000 parts also run in the US, six weeks ahead of schedule. In parallel, the second tool, a steel production tool, would be made by the Quickparts factory in China with qualified, focused engineers tracking the tool every day. The production tool would be ready in eight weeks, and 400,000 parts would then be run in China to meet the Pal-Mart deadline of March 1st. On a hunch that Todd might crash his job, Johnny then asked Helen to ship the production tool back to a local molder in the US where additional parts could be run if needed. Feeling extra witty, he reminded her that opportunity is where preparedness meets someone else's bad luck. Helen laughed out loud and added, "Oh Johnny, your production approach is uncanny. It sounds just like our Hybrid Manufacturing Strategy that sees the world as a superstore—mix and match to get exactly what you need; build the production tool in China and move it to the US for processing." Johnny agreed with her that Pal-Mart would love him for delivering 100,000 bunny radios early, while making sure his production tool would check out in time to deliver the rest at a super low cost. He also agreed with her that only a real Marlboro man can "grab the globe" and think this way. "Yes, Helen," he grinned, "the world is *our* oyster." Those must have been the magic words to unlock Helen because she quickly ensured him not only VIP treatment during his China factory visit but also a Deluxe Boondoggle Enhancement (DBE).

Johnny couldn't wait to check out his T1 parts in the Far East. He envisioned the historical importance of "East meeting West" in manufacturing. More importantly, he imagined endless mountains of steaming rice and Peking Duck on his expense account. He couldn't wait to try the deadly fire water "By-jee-oh" liquor that tastes like rubbing alcohol, and of course, the eggs that had fermented for 30 years deep in the Earth. If it meant being a gracious goodwill ambassador, Johnny was even willing to taste deep-fried chicken feet. Days before he left, he dreamed of mysterious Chi Gong Masters, the old Chinese guys who break people in half just by thinking about it. He marked the map for a visit to the ancient Taoist Mountain Shrine. As usual, Johnny was open to learning as much as he could about everything.

But leaving the great US for the first time ever, Johnny suddenly felt unapologetically homesick as he crossed the threshold into the East. He first saw China through spontaneous tears; a strange, tessellated world made of running watercolors and mysterious ancient feelings. The hectic streets of swirling color amid diesel fumes and pungent open-air cooking made a potent impression on Johnny's soul. After a day in Shenzhen, Johnny was in culture shock. Everything seemed too loud. He wrote Sally Savealot in Procurement a postcard that simply said, "There's no God here. But I am popular."

On the street corner, local ladies mistook Johnny for dead American movie stars, yelling out at him: "Hey, Cary Grant," "Hey, Gary Cooper," and the most famous cowboy ever, "Johnny Wayne." At street corners, Johnny peeled the women off his arms and walked away strong and steady. No matter where he ate, his soul felt off kilter. He tapped his work boot heels together three times and uttered, "There's no place like home." But nothing happened. He was still there with 1.3 billion of his closest friends, all hunched over cell phones and yakking with a fury he had never known.

Arriving at the Quickparts factory in Panyu, Johnny suddenly felt better. It was a week before the important trial T1 date, and Johnny had a Tooling Manager, Project Manager, Operations Engineer, and the China Project Manager all by his side. He felt a thousand percent wonderful about getting his project done right by professional engineers. He felt secure when the staff showed him their Engineering Change Order (ECO) system for tracking all changes to his production tool. Johnny could see that all issues and design review information had been tracked and communicated in detail. His team assured him that if anything urgent came up they would call him on his cell phone, amply supported by China Telecom.

This factory had taken all the fear out of his production tooling project. Besides, according to an email from Helen Helpalot in the US, his first 100,000 parts from the successful low-volume injection mold had just been delivered to a very happy Pal-Mart. So Johnny could afford to take a few days off to sightsee with his lovely translator, known as "月亮珍珠運氣公主好," translated as "Moon Pearl," or her American name "Becky." True to his nature, Johnny used her ethereal name.

Moon Pearl didn't even wince as their cab driver sped in front of a rickety 18-wheeler whose driver was out of his mind. Johnny gripped the seat as hard as he could and yelled at the driver to slow down, but to no avail. His cab was a bucking bronco. Johnny closed his eyes, said a little prayer, and then let go of everything familiar. If he ended up dead in the middle of China with no God, no church, no friends, and no next of kin to call from his cell phone, then it was just meant to be. By the time Johnny had finished the invitation list to his own funeral, the insane cab driver jumped the curb through a throng of shoppers and screeched to a halt in front of a "Western" hotel. With a flourish, the cab driver shouted and waved his arms, pushing everyone aside to make way for the new king in town, Johnny Quickparts.

By the time Johnny got to his room, he was ready for a beer and a little personalized karaoke. He couldn't read the Mandarin fine print on the complimentary packet of Chinese Liquid Condom by his bedside, but figured it said: "Only 50% reliable; gets rid of all the girls."

That night Johnny experienced his first "K-TV," the kind of mild "high-tech whoring" that even Granny Quickparts would approve of. In a livingroom-like bar, Johnny sat on a sofa and selected one girl out of 15 to hold his hand and sing with him for several hours. Johnny's girl did not speak much English but was very effective with sign language. Her gesture of tossing a drink down and yelling something like, "Gan-bay!" meant "Drink up!" in any language. She could also wail every syllable of Celine Dion's *Titanic* love song, "My Heart Will Go On." Under force, Johnny sang the song about 10 times before calling it a night. Late in the evening, he politely refused the "upgrade," whatever that was. Through a drunken buzz, he did retain some learning: the Mandarin word for America is pronounced "Mei Guo" which translates literally as "beautiful country." He also learned to ask for "mei" hotels and restaurants to end up at more Westernized places. Maybe China was not so scary after all.

Early in his project, Todd Hubris also stayed in China for two weeks and avoided a brush with culture altogether. He read his morning paper, *China Daily*, at McDonald's and ate lunch at KFC. He controlled his experience closely so that nothing new could ever disturb "Planet Todd." The factory he had selected was very clean and focused, and seemed to care only about his project. His entourage followed him everywhere, warmly calling him "President Hubris." He liked seeing his parts on the conference room table, and he liked having 20 people bow down to him. After several days of the royal treatment, Todd felt very secure about his project. He didn't know this great hospitality would fade as soon as he left the building. He didn't know the factory would drop his job immediately and give priority to the next visiting "President" from the US. Todd couldn't even imagine his project would be ignored for the next three weeks.

Todd returned to the US and reported to SCROO-U management that everything was hunky-dory in China. The machines were the same, the software was the same, and so he concluded that manufacturing in China was the very same as the US. Besides, he had proof of daily email status reports from the factory to show that the production tool was progressing well. His trial day, T1, would be on January 30, at which time, Todd would receive 10 trial parts via FedEx. "Piece of cake!" he yelled to his boss as he left early to make the local Mensa Mixer.

Meanwhile, Johnny was half a world away falling in love with Moon Pearl, his first-ever ka-ka-communist sweetheart. Waiting for his T1 parts in style, the enamored couple hiked up the misty mountain to a shrine where Johnny spent a moment in silence to honor the Jade Emperor of Heaven he had read about. He thought it tragic that Moon Pearl could not relate to prayer at all. A government system that "disses God" wasn't fair to her or anyone.

When January 30th rolled around, Johnny approved the trial parts for his bunny radio housing. The features looked good, especially the sharp ears and rounded tail. He sent an enthusiastic email to Helen Helpalot confirming an order to run 400,000 parts, then return his production tool to the US.

That afternoon Johnny said farewell to the super-affectionate Moon Pearl. He gave her fresh peach blossoms to ensure good luck in the coming New Year. When she begged him to stay till Dragon Month, which meant April, Johnny was floored at his growing international appeal. To break her spell, Johnny splashed his face with cold water in the men's room and repeated his basic values out loud with conviction: "God, Grandma, Plastics, America." He straightened himself and walked straight to his plane. Moon Pearl tearfully waved goodbye, yelling with an operatic style, "Goodbye forever, Johnny Wayne! You da man!"

Back home at last, Johnny deplaned onto his favorite terra firma. Fellow passengers watched as he prostrated himself and kissed the ground with a surge of raw passion for the US. He felt like lingering and cuddling the US but decided that impulse should wait. Back at Acme, Johnny found a number of insecure emails from Todd-centric and wondered what was up with this new wave of man-love from someone he didn't even like. Todd didn't say anything about his T1 parts, which led Johnny to believe he didn't have any. With all the godless colors of the East still swirling inside him, Johnny raced to Waffle House for the All-Star Breakfast for Heroes with his hash browns "scattered, smothered, and covered." He suddenly understood exactly what a culture gap is—when there is no equivalent for the language to carry. He knew that his communist honey would be completely baffled if asked to translate his breakfast wishes in Mandarin.

Down the street at SCROO-U, Todd sat, day after day, waiting for a FedEx package of T1 parts. He emailed his factory in China for a tracking number but he got no response. He told his boss not to worry, "Todd's in charge." The next day, still no parts, no email response, and no phone call. Todd started feeling queasy, but put on a brave face at the staff meetings. He gave in to unusual, obsessive urges to wear black clothing and sweep the floors. His co-workers noted that perhaps the stress of China was getting to him.

It was February 5th before Todd actually heard from his Chinese factory. The factory manager profusely apologized and said that his government had turned off electrical power for awhile. "But we are back on now and will most definitely ship your T1 parts today! No problem!" the manager blared through the fuzzy-sounding speaker phone.

Despite regular assuring emails, Todd still had no parts a few days later. His frustration mushroomed as he sorted through lame excuses and lies from legitimate-sounding reasons for the stall. A few days later, his FedEx box arrived. Much to his horror, his trial bunny radios looked more like puppies, and they weren't purple, they were red. The ears were warped and the feet had slightly cloven hooves. A critical feature, the puffball tail, was missing. Todd's voice cracked when he showed his boss "Project Bunny." On speaker phone, the Chinese factory manager said, once again, they would fix everything, "No problem." He promised new bunnies, better bunnies. Purple bunnies would be remade and sent in three days for approval, "No problem."

On February 15th, Todd received the second trial bunny radio housings, this time purple, with long ears and tail, but they still had cloven hooves. To expedite things, he decided to order 100,000 of these be-deviled bunnies anyway and hire manual assemblers in the US to file down the hooves. Todd approved the tool and ordered 100,000 parts to be run. He then went searching for a US machinist to do this additional manual work. Todd quickly found that he was going to spend all the money he thought he was saving by going to China in the first place. He felt the heaviness of adding to his problems, rather than solving them. Meanwhile, Johnny was well on his way to successfully delivering large lots of bunny radios to Pal-Mart's warehouse a week earlier than the deadline.

Several days later, the China factory called and woke Todd up in the middle of the night saying that the molder did not have the required purple plastic pellets. The required GE material was on backorder for six weeks. Meanwhile, Johnny was sleeping like a baby because he had ordered *his* material at project startup.

The next day Todd called a US molder friend to purchase enough pellets for his project and have them shipped to China for a mere fortune. That afternoon, Todd got an email assuring him that the factory would start shooting parts just as soon as the factory received his plastic pellets. The email ended with, "No problem." Todd breathed a sigh of relief, thinking that he could still deliver a partial order to Pal-Mart by Feb 28th.

When Todd finally confirmed that his plastic pellets had arrived at the factory in China, SCROO-U management breathed a collective sigh of relief. On speaker phone, the China factory manager announced, "No problem. We will run these parts as soon as our workers come back from

New Year's celebration. The factory re-opens in a couple of weeks." Todd had learned the hard way that Chinese New Year is more like a socialist workers' two-week vacation. It happens not on January 1st but according to an ever-shifting lunar calendar or government whim. Todd went into a screaming rant, until he finally blew his stack. Like Krakatoa, Vesuvius, and Kilauea, he erupted for 30 minutes non-stop until, finally, Todd Hubris was a weeping, hot blob in the fetal position on the conference room table. He could not persuade 1.3 billion people to give up their holiday through a scratchy speaker phone, thousands of miles away. In front of SROO-U management, his China factory partner, and the world, Todd Hubris had officially "lost face."

Needless to say, SCROO-U spent thousands of dollars without delivering a single thing and still missed Pal-Mart's deadline. Pal-Mart cancelled the contract and yanked the R&D budget in a matter of hours. Todd took his pink slip to the highest bridge and jumped. On the way down, he thought about everything he would change if he got to live.

Fortunately, Todd lived and learned, and now encourages everyone to think about the consequences of their decisions early in the game. He later went on to become a cashier at Dunkin' Donuts, after a short stint at Starbucks as a barista. However, SCROO-U Unlimited went bankrupt, pensions were lost, divorces were finalized, and other management jumped off bridges—all because of one project leader with enormous blind spots. Todd Hubris, a great engineer, pretty much did "know it all," but he became too content within a closed universe.

Several weeks after his return, Johnny was publicly made a hero at the Annual Cake Walk by Acme Design's rarely seen President. While both Johnny and Bob Overrun got huge raises, Johnny got a corner office and surpassed his boss with grace and humility. Not at all surprised to hear about Todd's company disaster, Johnny said it proved once again that parts really are the center of the universe.

---

## ▒▒ Boring but Necessary—LVIM Design Issues and Limitations

Plastic injection molding is challenging. As a discipline, it offers a degree of unpredictability. No matter how well you design your part, the LVIM process will add other features, errors, and effects that you do not want. These tool design issues are the consequence of the innate limitations in LVIM process. Finding that your trial plastic part has annoying anomalies is part of the high price of producing thousands of the parts fast. **Poor design of your LVIM can result in costly rework.** While engineers tend to think some issues are the manufacturer's call, it's best to communicate with all collaborators early in the process and design-in those decisions on the front end, especially draft. Key elements contributing to excellent plastic design include the following.

### Plan the Parting Line Design

The parting line happens wherever the halves of the mold come together and mate. This is where the part halves will meet to form a tighter bond. While this is not part of the design, the process will add a feature to your part and you must be prepared to use that feature to your benefit. One of the issues with parting lines are that they can appear in places that are visible to the user, which may be ugly. They can also affect mating places of the part with other things in the product, or over time can affect the overall tolerances of the part. While you will have parting lines, the engineer needs to design the part to incorporate the parting line into his design to use it or prevent it from affecting the part's functionality.

## *Draft*

Draft is the required slant or slope of the walls of the part that touch the sides of the tool. Proper draft allows the part to disengage quickly from the mold when the process is complete. Most engineers struggle with draft because they don't understand how the part will actually be molded, or they can't get their CAD software to work with the addition of draft. For such a simple feature, it can be a real nightmare in the CAD world to get draft on the surfaces in the CAD model without the model becoming highly inflexible. This user-nightmare is related to the complicated mathematics required for CAD surfaces.

It is common for the engineer to avoid draft all together and push it off onto the manufacturer. This is fine except when you let others control your destiny, you get your destiny controlled. The manufacturer may apply a bigger angle of draft on walls that are critical to your design and thus prevent it from functioning correctly. The manufacturer may also inadvertently prevent mating parts from mating with an increase in angle. The effect of draft is a function of the length of the affected surface and the angle of change. Letting a manufacturer change draft could result in features being bigger or smaller by significant amounts (tenths of inches).

## *Ejector Pins*

Ejector pins make features that are remnants of the process. These features appear wherever ejector pins were located, strategically placed to eject the part out of the mold when finished. While typically they are designed to be flush with the surface, ejector pins can be under the surface or may need to be located on a critical feature that can cause tolerance or interference issues. As the designer, you have little control over the placement of the ejectors; however, if you understand the process of injection molding then you can be sure to indicate ejector pin locations and communicate those to the mold maker.

## Materials for Parts

It's all about the materials. At the end of the day, you are using injection molding to make use of great materials that will suit the needs of your parts. The LVIM process is very amenable to these materials and ensures the efficiencies needed to replicate your part quickly. However, the designer must be aware of what he expects from the part and materials, as both are interrelated.

The key issue with the material is viscosity, or how easy it will flow in the mold. The designer must select material that will flow in all parts of the mold before the material cools back to its solid state, or he must design the part such that material can easily reach all areas of the part. If the part has thin features, like cooling fins, and the material is very viscous, then it is likely the tips of the fins will not form completely. However, if the material was less viscous, then the fins would have no problem forming.

Also remember that materials respond differently in the LVIM environment. For example, running the same mold in both Polycarbonate (PC) and acrylic will give you two different kinds of parts because the melt flow and shrink factor of the materials are dramatically different. One automotive company had to pay additional money to convert the mold to run in acrylic because the mold was originally built for PC.

Material selection must always be feasible for the part design. Be sure to choose a material that lends itself to successful molding of your part design. A material that has a high-warp tendency is not good for product applications requiring a strict flatness specification. Tool modifications may be necessary to compensate for material or part design discrepancies.

> **QuickTip:** Over 40,000 thermoplastic materials are currently available to use for parts made with LVIM.

## Aluminum for Tooling

Most LVIMs are typically made of aluminum, and aluminum has limitations.

Compared to steel, it does not offer longevity or consistent production quality. Aluminum is not good for molds that run under higher temperature requirements. It can have challenges with cosmetic finishes or smooth tooling surfaces provided by harder material tooling. Since aluminum tools are soft, they can be machined and polished much faster than hardened steel production tools. Aluminum can fail when used for tooling to build electrical connector parts because it can't form long, thin pieces.

One electrical company tried to use aluminum tools to produce connector parts that needed steel inserts. Aluminum was not sufficient for the tooling needs, and the final price approached the high cost of making a production tool.

## Tolerances

As with all manufacturing, tolerances exist in LVIM. The standard tolerance is ± 0.005 inch (five thousandths). While we can design the perfect part with the perfect dimensions, we are unable to ultimately produce this perfection. When you are designing with a melted material injected into a void to solidify, maintaining perfection is nearly impossible. The designer must be aware of these variabilities in the design and account for them in the functionality of the part. It is very common for great designs to fail because they cannot be made close enough to perfection to work. This requires that other parts get changed to accommodate the imperfection or the product will have severe issues.

In the tooling world, the prediction of these tolerances and how they are made is somewhat like artistic guesswork since the geometry of the part, material, tooling material, pressures, and many other variables affect the output. As the material transitions from a solid pellet to a liquid flow to the solid shape of the part, shrinkage occurs, which can affect the tolerances of the parts. The amount of shrinkage is a function of the materials used.

## Lead Times

LVIM typically takes two to six weeks, depending on complexity. A contributor to the short lead times is the use of CAD data to drive mold design and eliminate paper drawings to build the mold.

## ⠿ Saving Time, Saving Money—Saving Thousands with LVIM

### *How Do I Save Money Using LVIM?*

In manufacturing, **almost everything you can do to save time will save you money.** For starters, build a mold that will support the quantity of parts needed. Use LVIM only when you need a low quantity of parts and are sure that you won't need the capability of a production tool. Inaccurate forecasts of low production needs cause the per piece price to exceed market allowances, thus creating a need for automatic production tooling. If you are going to make 10,000 parts at a time, your service provider should be able to add automatic slides to reduce cycle time and operator requirements which can be costly in high volumes.

Be sure to understand the best use of LVIM over CU for certain designs, since for many designs, LVIM is deemed more cost effective than CU only after running 50 parts or so. Consider producing parts in large batches for use over longer periods of time. In other words, consider running the total parts needed for one year to keep your price per part lower.

There are many LVIM design considerations. Be sure to learn the limitations of the process—radii, tolerances, feature size, and wall thickness—and the consequences they have on design. Design features with appropriate radii for machining can help you avoid the cost of additional EDM work. Keep parts as simple as possible to eliminate the need for hand loads and additional tooling costs. Designing with cutouts or windows for snap features and undercuts means easy access for manual tools down the line. Avoid making design changes and concessions before production.

When designing large tools, use LVIM to create a completely CNC-machineable part to reduce the need for EDM, a time-consuming and expensive process. Material removal is much faster using CNC. If a part design can allow for complete CNC machining, tooling can be delivered in a matter of several days, depending on complexity.

Always verify that your part is capable of being injection molded using Design for Manufacturability (DFM) rules. This will prevent you from investing thousands of dollars in a production tool that would need excessive modifications in order to actually produce the part you design. Use services to test the design before graduating to tooling.

Informed engineers always verify their design in LVIM to save money. One medical equipment company currently spends $300,000 per year in SL and then moves into LVIM to verify the part design. Integrating production suppliers in the LVIM process helps everyone with the learning curve of manufacturing a part. An automotive company uses LVIM to verify the part design for their customer. Purchasing LVIMs after the release of the production order gets the customer to sign off early on the parts, before submitting production tooling parts.

Troubleshooting the design with RP prior to making a tool saves on costly rework where small changes to mating or function are needed. Reviewing first article parts completely will also catch defects and prevent costly production of unacceptable parts. Be sure to sample the mold in various materials, colors, and textures before committing to a run.

Another money saver with LVIM is that molding issues can be worked out in single cavity versus multiple cavity tools, while proving out part designs for the function of the application. After the part has been qualified with LVIM, companies are able to produce better, automatic production tooling at a more cost-effective means. LVIM also provides the savings of incredibly compressed timelines of two to four weeks, while creating an automatic tool (eight to twelve weeks) would hinder getting the product to market and result in lost revenue to the customer. It's important to incorporate parts into the production cycle while the production tooling is being built and coming online for the customer, saving sales that would otherwise be lost during the tooling process.

Requesting sample LVIM parts can be another money saver. Using functional samples in the assembly setup while your final production tooling is being made can save time when the production parts finally begin to arrive. Marketing samples will also get solid market feedback on a design prior to costly production tooling. Samples are useful in packaging studies as well.

The most controllable money-saver is clearly defining and communicating all part and project specifications up front in the LVIM process. This includes selecting the best end-use material for the design and use of the part, supplying current CAD data for quoting or production, and providing information on the intended use of the parts produced and expected results. Of course, always provide the final file versions at time of order, knowing that the clock cannot start on your job until all data is received.

### How Do I Waste Money in the LVIM Environment?

If you do everything you can to save time and save money in your LVIM process, you can avoid these wasteful scenarios.

Many tools are built in error, either due to sending the wrong revision or to hoping that non-conducive geometries might somehow work. Parts with very thick sections undergo a significant shrinkage defect to the entire diameter of the part, causing failure. Expensive tooling changes and engineering change orders (ECOs) are often required to compensate for part design issues.

Design decisions can also waste money. Avoid designing parts that have side actions, and watch out for designs that need multiple threaded inserts. Additional costs hide in parts designed with many side actions. With these parts, customers often expect a much lower price than what they actually get.

Avoid changing design or materials in the middle of a job to save on complex rework time. One automotive company tried to switch from PC to acrylic material after tool completion and was unable to produce acceptable parts in acrylic. The company had already sold orders for both materials and was faced with having to provide a product made of only one material. Another company provided a part design that was not conducive to good molding. An inner undercut did not work well when ejected from the mold. The company had to change the design and adjust the tool, losing valuable time on an already tight lead time. Also, remember that it's very expensive to change a mold that requires the addition of material in order to create a new feature. Product engineers are faced with designing around inserts when a "boss" and self-tapping screw would be better for cost in the short run and much better for part pricing in the long run.

Wrong expectations are the biggest money waster. Don't expect that a tool with multiple actions or inserts can be made on a shorter LVIM timeline when you really need a 12-week schedule to produce an automatic tool.

## How Do I Save Time Using LVIM?

The best time-saver is to make sure that your part is designed for plastic injection molding. Tool build lead times can be significantly reduced if part designs do not require EDM or side actions. Producing single cavity LVIMs for development purposes allows you to do tool modifications quickly. It requires much less time to revise one cavity than a multi-cavity mold.

Always allow testing of parts before releasing production. Choose the best process—LVIM saves more time than CU after only about 50 parts. Schedule your order well in advance of your deadline. Plan ahead to prevent taking shortcuts, which ultimately do not save time.

## How Do I Waste Time in the LVIM Environment?

Administrative flubs are hidden time-eaters. Be sure to issue a valid purchase order with the project start, and prepare your finance team to pay the first 50% deposit to get the project started. Always reply quickly to your service provider's request for design approval or concessions. **Start a project with the final correct files, knowing that the clock does not start ticking until a purchase order (PO) and the correct version of design are with your service provider.** Have realistic expectations about tolerances—it is not uncommon for companies to expect a tolerance of ± 0.001 (one thousandth) when aluminum tooling tolerance cannot meet this.

## :: The Keys to a Brand New Amphibious DUKW

By now you know that LVIM is a manufacturing method that creates mostly aluminum tooling or injection molds for producing short runs of up to 50,000 parts. You've explored China with Johnny to witness an aggressive product development race, and you've been inspired to try LVIM as a "bridge tool." You've felt the technological excitement of seeing how LVIM can compress production time and get parts for evaluation early in the process. You understand that LVIM provides powerful insurance on very costly projects. You've seen the world of manufacturing turned upside down and you now understand the winning strategy is called the Hybrid Manufacturing Solution. Enjoy the chaos.

As Johnny would say,

**"Entropy Rules!"**

# Chapter 7

# Strategies for Production Plastics

## THE ALPHA AND OMEGA

*"In creation, nothing happens until God
makes light and you make parts. Parts
are the center of the universe."*

## :: From Process to Production—Pregnant with Possibility

Developing a product is like gestating a fetus, slowly bringing invisible inspiration into form and matter. We know the thrill of watching the phases of change as the invisible creative powers inside of us, with the help of Mother Nature or additive fabrication, miraculously turn that "nothing" into something magnificent.

The next and last phase of the product development cycle is called production. You are now ready to make the "final push" from incubation to an actual product that stands on its own in the world. As a product developer you may feel a little lost in this netherworld between phases. You are finished with all of the pre-production processes but are just getting started with production, where product "life" begins. After production, all you will need is a little elbow grease and magic to assemble a pile of parts into a market-dominating product. It's time to deliver!

**QuickTip:** First comes Tooling, then comes Parts, before you can sell cool stuff at your local Wal-Mart.

The previous chapters described the critical steps in processes such as Stereolithography (SL), Selective Laser Sintering (SLS), Fused Deposition Modeling (FDM), Cast Urethane (CU), and Computer Numerically Controlled (CNC) machining for turning a design into a physical part. Low-Volume Injection Molding (LVIM) processes were also offered as an entrée to production in lesser quantities. In the production phase, you now prepare to deliver a certain number of widgets to your customer. You need to know where to get the "real thing" made as economically and quickly as practical. The focus now turns toward finding the best strategy to make this final push for hundreds of thousands of plastic parts. You are faced with the most important decision in this process: Who will be the "mother" of my widgets? Who can produce my parts quickly and reliably to get them on the shelf?

## ▥ Getting to the Wal-Mart Shelf—Are We There Yet?

Since the beginning of the Industrial Revolution, the way companies manufacture their products has changed significantly. A clockmaker in the eighteenth century would make all of his own parts, assemble the clock, and sell it. But today, because of the overhead cost associated with manufacturing operations, it is typical for many companies to outsource the production of individual components as well as the assembly of those parts, then stick their own label on the final product. Whatever it takes, companies want products, at the lowest cost possible, to be on the shelf "yesterday."

For executives who care about global competition, the winning attitude is to accept that the world is changing. Therefore, we must "change or die." You are in business to get parts quickly so they can be assembled, packaged, and shipped. If you have already gone through the processes described earlier in this book, you have invested thousands of hours and dollars to get an idea to reality. Therefore, business managers and buyers need an alternative strategy that will perform better than a familiar default strategy. Typical default strategies are simple: either use China only for both tooling and parts, or use the US only for both tooling and parts. After all, parts are the end-all and be-all of the manufacturing base, a product developer's reason for being, the Alpha and Omega of product development, and our connection to consumers, sales, and revenue.

At a quickly increasing rate, the current trend is for companies to reduce their workforce and manufacturing capabilities. In the early '80s many companies began outsourcing their own projects to US manufacturers in efforts to make their operations "lean and mean." While "just outsourcing" worked well throughout the '90s, the competition-driven demand for lower costs in manufacturing is now shifting the outsourcing requirements to include using non-US manufacturers. If you are a product developer, you probably already have a mandate to "make it happen" in China. As a result, China has become the manufacturer for the world and is forcing the US manufacturers to reassess their future.

### Where China Races Ahead of the US

Manufacturing is the critical economic sector in American society. Without the manufacturing of parts, products would be unavailable to the market and our economy would fall. Today, the many challenges of being a manufacturer have already killed a big part of the US manufacturing base. Two key challenges for any manufacturer include capital equipment required for growth and flexible human resources. These two resources, greatly simplified here, are responsible for the demise of the US manufacturing base and the reason China has become the dominant provider of manufacturing services to the world.

#### • Capital Equipment

In order for a manufacturing company to produce parts, it needs equipment that can produce the parts. For the manufacturer to grow, it must buy more equipment to expand its capacity. To get more equipment requires more available, usable cash. If a company is unable to produce a positive cash flow to support new equipment acquisition, then it will be unable to reach the profit zone. The ebb and flow of revenue and cash flow can make it very challenging for a manufacturer to add more equipment, required for growth, especially when future revenue is but a dream—a dangerous cycle for the manufacturing industry.

Unlike the North American manufacturer, the Chinese manufacturer is able to overcome capital equipment constraints because the Chinese government can subsidize their business and provide more than enough equipment at a low cost to the manufacturer to support expanding growth. This approach has created thousands of shops throughout China and has provided the opportunity of a great future for many.

Another, less obvious advantage of the Chinese manufacturer is with software. It requires sophisticated and expensive software to operate the equipment. In China, it is possible to buy $100,000 worth of software for the cost of a blank CD. You can visit some companies with thousands of employees and they will not have a current software license for much of the software. Interestingly enough, there is no remorse or guilt with the use of illegal software. Some business managers rationalize the use of pirated software by claiming that the cost of the software is so high, most in China could never afford to buy it. So, if they can never actually purchase it, then it is okay to use illegal versions to operate their business. Interestingly, there is a unique situation that is building in this area. As the major software companies have gotten smarter, they are now locking software to prevent piracy. This is becoming prevalent in many of the shops in China where they have been unable to upgrade to the latest versions. In time, they will have to figure out how to get to the current version in order to serve their customers. Also, as China continues to develop their own technology, in software and innovative products, they will become much more sensitive to others using it without permission. So, in the next five years, there will be a shift occurring that may have an equalizing effect on global manufacturing. Regardless of the ethical opinions of the advantages provided, they are real and have an impact in a global economy in which the world continues to flatten.

### • Flexible Human Resources

The second key aspect of manufacturing is the need for flexible human resources. As projects move through the organization, the need for people fluctuates. If an organization maintains more than the required human resources, it reduces its positive cash flow due to organizational waste. This reduces working capital opportunities, thus inhibiting the growth of the business. If an organization maintains less than the required human resources, the company struggles to satisfy the requirements of the current projects and risks failure, which prevents future projects and again inhibits growth. In the US, the opportunities to use flexible human resources are continually being eliminated while employee regulations increase. With a growing economy, the need for flexible human resources continues to expand.

Chinese manufacturers are able to overcome the flexible human resource constraints since there are literally millions of workers who are willing to work for minimal wages. It is not uncommon for a factory worker to work 12 hours a day, 7 days a week, and earn $100 per month, plus food, clothing, and shelter. Chinese workers are extremely grateful for any kind of work. Most have never heard of employee rights, wage minimums related to the cost of living, basic human rights, vacation, sick time, or any employment protections held sacred in the US. In time, these advantages will change as the Western world continues to influence the Chinese by making them aware of a different paradigm. Also, as China continues to accept their role as a world leader, the expectations of the world will put forth incredible pressure for them to change.

China's strengths provide a strong competitive edge to continue to grow and maintain their status as the leading manufacturer of the world. However, the US manufacturer has an opportunity to continue to transform into a segment that can profitably operate in the new economy, most likely through low-volume manufacturing. As discussed in the previous chapter, low-volume manufacturing does not have human resource constraints like mass manufacturing. Low-volume manufacturing allows the manufacturer to specialize in an area, such as tight tolerances or fast lead times.

### *Neighborhood Watch*

It's a tiny world now, made of shifting sand. China moved in across the street, next door to India. As new neighbors, we are rapidly grasping to learn each other's ways and expectations. The pot of global currency we share is swirling around in new combinations. Each morning we wake up to find out what we lost while we were sleeping. After breakfast, we find courage, get our legs back on, and go out to war on a dizzying terrain for daily bread once again. Needless to say, the manufacturing neighborhood is a bit tense. We suspect that someone has "moved the cheese"—*all* of it. But US manufacturers have not yet put on their running shoes. They keep going back to the empty room to wait for their cheese to show up, but it doesn't! That cheese is "gone-gone."

Product developers have expanded to using a global supply chain for manufacturing. Companies now implement systems to acquire parts from all over the world to access the lowest cost from China, India, and Vietnam. You can look at this new manufacturing world as an opportunity or threat, but in the end, businesses will either adapt or die. It's time to revise our vision and see with new eyes. We need to understand the world as a whole, not just our little corner of it. It's time we navigate through the global manufacturing world as a colorful, open-air bazaar of choices, fraught with perils and pearls, rip-offs and rewards. Like tourists in a strange country, we all want a quick lay of the land. We all want to know what's SWOT (strength, weakness, opportunity, threat) and what's not.

## Protecting Your Manufacturing Future

Trade journals are full of differing stories attempting to quantify the devastating manufacturing losses in the US over the last 10 years. What we know for sure is that US manufacturing is at its lowest output since 1958. However, the US can overcome these hefty losses if manufacturers focus on and invest in only their core competencies, then spend dollars in China for select manufacturing processes that have been carefully analyzed. A powerful strategy, the Hybrid Manufacturing Solution, leverages the strengths of the Chinese manufacturer and serves the needs of the US product developer. There's nothing fancy about it: **make your tools in China, make your parts in the US.** While it sounds simple, the China maneuver is way more challenging than a few phone calls. In fact, you have to change your perception to really succeed on this expanded playing field.

Building on the previous product development processes we've discussed, the Hybrid Manufacturing Solution thesis is that 95% of a product can be developed in the US, with only 5% of the labor needs for mold making going to China. By building an efficient organization that can reliably provide mass-manufactured parts to customers, US manufacturers will be able to support their own economic growth by selling more product and maintaining profit margins. Instead of building factories, the US provider can free up cash to add systems and people to better serve the customer.

If you are in the business of developing new products, then you are in the business of accessing the best resources in the world. Today, China is an important part of this strategy. Before discussing the Hybrid Manufacturing Solution in detail, along with what *not* to do, a closer look at China will provide a context for understanding how and why the Hybrid Manufacturing Solution is the only international strategy that consistently works.

## :: China—The Manufacturing Capital of the World Now

**China now touches your life every day,** whether you want it to or not. Go to any store and read the "made in" tags. China's "commi-capitalism" spans wide: from manufacturing the products you use, to the continuous accumulation and holding of US currency, to a significant contribution to global warming with uncontrolled pollution. In many ways, China is as unrestrained and aggressive today as the US. The scary part is that China is as entrepreneurial now as the US was back in the 1900s—scary, especially when innovation in the US continues to dramatically decline. Clearly, China is the manufacturing capital of the world today.

This position is well deserved by the hard work, dedication and focus on changing their world so they can be in a position to change the world. China is probably the only country that can transform so quickly and continue to take their place as a global powerhouse. This does not mean that they are doing everything right or in the best interest of others, but it does mean that China deserves the respect of being a great resource for manufacturing in the world today.

While China is now the natural solution for manufacturing, we have to test the assumption that China is the perfect solution for every project. Despite the risks, it is safe and actually politically correct within the walls of product development to get your parts and tools made offshore, but there are plenty of challenges. Going to China does not automatically keep your job secure. Product developers need to understand that China is most definitely *not* the US, and it's not always cheaper. Scrutinize any magical thinking on your part and prepare to overcome hidden obstacles lurking ahead. As they say about China, "The rules are different here."

### Empowering the Red Sun—A Motivator to Succeed

> *"Socialism must be developed in China, and the route toward such an end is a democratic revolution, which will enable socialist and communist consolidation over a length of time. It is also important to unite with the middle peasants, and educate them on the failings of capitalism."*
> —Chairman Mao Tse-Tung

It's a strange time in American history. As the democratic superpower, we find ourselves empowering the enemy, communism, albeit a restyled, capitalist brand. Stuffed full of US dollars, China is growing so powerful that it can challenge the US military or economy anytime it chooses. Furthermore, by using China for manufacturing, Americans are empowering a country that blows up missiles in space and heavily pollutes the space belt, not to mention the planet's oceans and atmosphere.

The point is not to pass judgment on the actions, nor to demean or ridicule China. Through hard work, dedication and determination they have overcome decades of challenges and hardships. It is a unique era in history when the world is so close and those that could be enemies can also be partners in business. This is the relationship between the US and China, and a situation that all should be aware of, and not fear. A relationship that should be intellectually challenged and appropriately managed for the good of all mankind.

Trading with China will have political ramifications as yet unseen. Imagine our future retail superstores bearing posters of Chairman Mao next to the "blue light" special. Or maybe, 20 years from now, US citizens will be pressured through twists and turns of policymaking to learn to speak Mandarin. In other words, this new relationship with China can impact our personal freedoms in the very near future, but no one is calling "the Red Sun" a threat. Americans love cheap goods, but is the emperor wearing any clothes?

Spending billions of dollars each year, the US buys friends like India to hold a balance of power, just in case. Meanwhile, Russia further empowers China to become even more powerful than the US. Americans have no assurances against a grave political threat from China in the future, so it's best to be aware of what we are empowering and diversify our global supply chain.

Many of us cannot even imagine the lack of freedom still found in a uniquely "communist capitalist" China today, where schools are still required to teach Marxism. Even though China opened up its markets some 30 years ago, you can still feel oppression in the air and see a lack of innovation in products. However, the power of product development has afforded China new status as an economic superpower and an important source of global manufacturing. China's very fast rise to power is due to free enterprise, a choice that shaped its social and business environment. Therefore, we in the US need to wake up fast. While China has a positive, heroic connotation of the dragon, we need to understand that the dragon is out of her cave and she has very long claws.

Education drives economic expansion. One of the great advantages China has is access to the US educational system, a resource we offer freely to global neighbors. Chinese students come to the US to study the history of what succeeded and what failed. This applied knowledge radically drives their economic expansion unlike any other country in the history of the world. In contrast, the USSR resisted free enterprise until its demise. While the Cold War pressures of the US brought about their demise, the USSR's lack of economic expansion from free markets allowed the final death blow. The point here is that **free enterprise nurtures the creativity of product development and unleashes powerful, positive change in the world.** We must use the growth and power of China as a motivator to succeed, not something to fear. China has done an extraordinary job of developing over the last 35 years. Regardless of politics, China should be commended for its economic success.

### SWOT and Sweat

The Chinese manufacturing sector is committed to success at all costs. Besides the fact that the average wage comparison of the US to China is still

**QuickTip:** Like electricity, China is neither good nor bad—it just is. Electricity can be used to light your way or put you to death.

20 to 1, China's labor pool is comprised of over a billion people with a fierce work ethic, a resource unfettered by regulations. The Chinese government also heavily invests in and subsidizes factories to produce parts and products. For these reasons, China can manufacture much less expensively than any other country in the world.

Alongside the US, many other countries have lost a huge part of their manufacturing sector to China. In fact, Chinese manufacturing is the fastest growing segment in the world. Evolving for the past 35 years and just now approaching critical mass, China will soon have total dominance over the global economy, thanks to an expanding infrastructure with reliable electric power and highway systems. Increasingly, large companies such as Wal-Mart, GE, and HP are also supporting this growth by demonstrating success with "Made in China" goods.

China offers innate risks and weaknesses to challenge the product developer. Manufacturing in China is not the same as the US—we approach quality, workmanship, and resource management very differently. For example, China solves problems by simply adding more people to fix something, regardless of their skill set. Needless to say, you need only one good driver to drive the cab. Even though the Chinese manufacturer may speak English (sort of), and use the same equipment and even the same software, there are still enormous differences to overcome. In the US, a product developer will send his project—problems and all—to the US manufacturer expecting them to "just make it work." In China, you get exactly what you send and sometimes worse.

The term "Chinese quality" is an oxymoron, the *perception* of quality in China is "If it works, it's good." Traditional systems and common sense structures for doing business are sorely lacking; innovation and creativity are missing in the offshore problem-solving department. If it's easier to ignore a problem, the China factory liaison may not even see a design issue or care to fix it. Offshore factories don't really seem to care about workmanship if the customer is thousands of miles away. The cultural need to "save face" seems to be triggered only if you are standing face-to-face with a China factory manager. In other words, you get better results if you hover on the factory floor. In China, workmanship pride is inversely proportional to the miles between you and their factory.

International companies need to anticipate additional risks of doing business in China. Primarily, governmental or political conflicts in a region, along with trade relations, can seriously affect your business. Fluctuations in currency can also impact your costs and profitability. Deploying marketing intelligence and diversifying your supply chain are basic risk management requirements.

The language barrier, even with English-speaking engineers in China, is still a serious obstacle. Experienced US-China entrepreneurs still underestimate the communication barrier—in China, English is not "English." It takes a newcomer a little while to sort out problems compounded by the language barrier. It's shocking to discover that China businesses typically do not have experienced management or basic business knowledge. "Chinglish" adds strain on communicating the specific technical details.

Pricing is another major weakness in China manufacturing and business. Still a capitalist teenager, China offers no discernible pricing strategies. This makes trading a challenge for everyone doing business in China. Product developers delivering to a third party have to be especially careful.

In China, the culture emphasizes the appearance of wanting to please the customer, even if the work lacks quality. The Chinese way is to give you, the customer, whatever you want, including the price you desire. However, based on your asking price, the manufacturer exerts control over how they get you what you want. This is the point where you can lose complete control of quality, schedule, and accountability. In China, it's easy to get seduced by a low price. Just remember, you would not want to buy a car for $50. If you do, you know that **your expectations need to match the price tag.** To say "You get what you pay for in China" is steeped in bitter truth. Since tooling prices can range anywhere from $1,000 to $100,000, be a smart shopper.

Despite numerous quirks, differences, and weaknesses, China means huge profits to product developers. Turn it around and ask, "How can I be successful in spite of China's challenges?"

### China—Not Just Another Supplier

For the past 100 years, American parts were made in the US and the makers of our parts communicated all nuances in English. Part changes were well documented within an established engineering change order (ECO) system. In today's global manufacturing world, our parts are made in far-flung places anywhere in the world by folks who do not speak English, do not track revisions, and may not even know or care where the US is located. This cultural disconnect can create problems that many business leaders haven't even imagined. Watch out for these dangerous gaps that can create new twists of technical miscommunication about your manufacturing needs.

Most US executives do not understand offshore challenges. They want to think of China as "just another supplier" in the Rolodex. The executive wants to believe that they can send their data to China and expect parts to arrive on the dock as promised, expecting the very same service they get from US manufacturers. If this were not true, he may argue, so many other companies would not be using China. This is a powerful argument in isolation of all the facts. However, successful companies are taking action and using protection to remain successful in China. The fact is that China is not just another supplier, and most Chinese manufacturers are neither as established nor sophisticated as US providers.

Many well-intentioned businesses fall into cracks between the two worlds. Many great products get lost in the void and end up as a sad story told too many times. The goal here is to provide "been there—done that—improved it" tips and strategies to help product development companies be successful in today's market.

## Fighting the Dragon

Every product development company has a mandate to produce parts in China to take advantage of the low costs. The challenge is that there are dramatic differences between the Chinese manufacturing system and the US manufacturing system, making this a tricky mandate. As you already know, it's not always cheaper and you can't always win in China, as evidenced by thousands of products that have failed there. To be successful in China, you have to fight for it.

A US company must be unrelenting to control its stake in China. Deploy trained, qualified personnel to live, yes, *live*, at the factory in China. Regular visits are not enough. Your team must be on-site to push production scheduling and quality control according to your own standards. You must have experienced program managers that understand how products are developed and are able to manage the project managers at the factories overseas. You must also have a very experienced operations manager in the factories to provide hands-on management of every aspect of daily operations. Create systems to track everything, and plan in detail to steer every phase of the process.

## Why US Companies Fail in China

US tool makers have been refining their process for over 100 years. China has been tool making for only 20 years. Typical American engineers think of tool making in phases of trials and texturing, going back and forth until the tool is right. In the US, a manufacturer makes the tool, then shoots parts as a trial run. If the parts do not meet its standards, the tool is tweaked and trial parts are re-run. This process continues until the parts are right. At that point, the manufacturer sends the tool to the texturing expert for a special texture such as "Moldtech." After that, they run more parts to recheck them and get ready for production. In China, however, there are currently limited ECO systems and little infrastructure for evolving tool improvements.

The most common mistake in dealing with China is to think that the supplier is really the supplier. From Guangzhou to Beijing, practically every business name is a façade, and it seems nothing is real. The US businessman spends his nights in a "Western" hotel sorting through the daily illusions, wondering about all the faces at the factory that day: who was the real decision maker, who was friend, who was family, and who was local mafia.

Other common mistakes result from not realizing how risk is reassigned when we are playing on someone else's turf. Doing business in China, you can expect strict payment terms, pre-payment requests, money-wiring hassles to a foreign country, outrageous shipping bills of $10,000 or more, a three-week minimum on shipping parts and tools, heavy import taxes, in addition to tariffs and duty holdups.

> **QuickTip:** Don't be surprised if you receive a container full of critical parts from China that instantly become paperweights due to poor quality.

An excellent example of how a project can fail comes from a key difference in the tooling mindset: China manufacturers "measure once, cut once," which means a lot of trial and error. To fix mistakes, the Chinese back up to reproduce the original work piece to cut once again. Sometimes the cut parts are welded back on, which introduces a quality issue. Texture on top of welded surfaces will look distorted and replicate badly because welding material properties affect the tooling material. In other words, a great deal of resources can be spent for a part that does not meet your requirements and is ultimately never delivered.

Many product development companies tell stories about how they went to Taiwan or China and failed, stating the key reason that business in Southeast Asia is not easy. Product developers can benefit from correcting a number of wrong assumptions and learning the truth. Namely, much of China does not hold the same standards for quality control and ECO management as the US; a continuous physical presence is absolutely essential in China for direct supervision of a project at the plant level; getting business done through the Chinese governmental system is very slow and difficult with many loopholes; and logistics management of getting a product to the US can be very complex and unimaginably expensive. Due to high shipping costs, there is usually no good option for resolving problems because it's not worth sending the parts back for rework.

Furthermore, many Chinese manufacturers are not trained in how to manage tasks and projects, so they are unable to produce reliable production schedules unless properly trained. Personnel in China tend to focus on only what is in front of them at the time without efficiently using resources which may be required to get all parts produced together for final assembly. It is not uncommon for China manufacturers to lack the technical capabilities their business requires, including computers and software, which are standard in the US.

All of these conditions can make business endeavors in China impractical for the naïve and inexperienced. Entrepreneurs need to learn to eliminate these risks by using their own or someone else's expertise, resources, and systems to efficiently provide mass-manufactured custom parts and products. The US manufacturers and product developers that implement the Hybrid Manufacturing

Solution will be able to leverage these strengths to provide a great service to US businesses and continue to aid in the growth of a global economy. In a market where problems are perceived as opportunity, experienced companies exist to provide you with services to make your Asian journey smooth and straight.

## Watch Out if Your Company is Less Than $250M

Interestingly enough, the answers to many questions about doing business in China greatly depend on the size of the company involved. If your company is greater than approximately $250M, then business in China will be a bit easier for you. At that level, your company already has the resources, expertise, and mandate to be in China, or anywhere else in the world, to make it work. If yours is a forward-thinking company, you have probably been in China for years with an established team and well-known suppliers. Other than managing the many day-to-day challenges with manufacturing in China, your company is probably doing well. Your biggest challenge is most likely the lack of real and accurate information about the regulatory, business, and political pulse in China. Faster marketing intelligence is currently coming available through internet-based technologies with systems that truly integrate the manufacturing world into a semi-homogenous system that is user-friendly to pioneers looking eastward.

Larger companies need to evaluate whether they have the right procurement resources and experience to manage China, as well as sufficient engineering resources to put in China. They need to know that their focus is on product development only and not on manufacturing, the reason for outsourcing. Larger companies need to name transition points with an updated vision of executive management, which can be a challenge if they still have executive management from when the company was much smaller. Another concern is upgrading key personnel who may lack experience or knowledge required to grow the company to a new level.

In contrast, if your company is less than approximately $250M in revenue, then things are bit tougher. Companies this size or smaller have a serious need to reduce manufacturing costs using China manufacturing, but they do not have the resources to directly manage the system successfully. The purchasing agents are typically inexperienced in non-domestic manufacturing and the engineering teams are too busy working on their next project. A smaller company hopes to send projects to China and hopes their parts come back on time and within specification. Unfortunately, **"hope" is not a successful strategy and it costs companies unnecessarily.**

### A Very Tall Tale—Johnny Quickparts Gets Promoted

With Johnny Quickparts at the helm of Acme Design Corporation, success was in the air. His wins in China were stacking up. Much to his own surprise, he was quickly becoming an icon in manufacturing, a rock star of plastics. Johnny's years of silent, persistent study had finally paid off. He found himself lunching with big shots from around the world, many of whom wanted to adopt the kind-hearted genius. The founding father of Pal-Mart, a sweet old man from Arkansas, was so impressed with Johnny that he personally put Acme back on the map.

Even though Granny Quickparts didn't technically understand Johnny's international world of part making, she understood the ultimate Product Developer, found in Genesis, very well. To celebrate Johnny's promotion, she gave him a plaque for his new office that simply read:

*"For it was You who formed my inward parts;*
*You knit me together in my mother's womb."*
—PSALMS 139:13

Johnny had finally earned the engineer's freedom to dream undisturbed in a nice corner.

As he squeezed a small rubber "world" in his hand, he could let his mind go wherever it wanted, without having to pretend he was working. He thought about his favorite novel *Brave New World* in which only 10 people ruled the world by eliminating most freedoms and perverting basic human values. Certainly the future society enjoyed massive wealth, pleasure, and mindless recreation, but they were all captive. People had forgotten their own creativeness.

Johnny let his new secretary pick up a stream of incoming calls as he pondered more important things like the invention of the very first wheel. He could only imagine what that scruffy Engineering Change Board must have looked like. Like a butterfly, his mind meandered even further back in time to the very first fire. More than anything else, Johnny wished he could have been there to see that satisfied customer's face.

Munching through a stack of Pringles, Johnny pondered his global family and its freedom to create, explore, experiment, and understand the spiraling cosmos. He knew that scientists, artists, and inventors thrive in the world of the imagination, living the question "What if?" He knew that without freedom, life's best creations—children, puppies, mangoes, trees, oceans, stars, and ultimately, the spirit—would somehow suffer.

As for the women in his life, Johnny put them on hold while he grew into a bigger self. It would take awhile for him to absorb his new level of success, and he had a feeling, whoever she was, the right gal would be there when he was ready. Johnny Quickparts was madly in love with work. He felt like God in the sandbox, shaping primal matter into products that hopefully would enhance life, give pleasure, or reduce pain—useful objects that would make ease, connection, and efficiency like the iPod, Dirt Devil, or Treo. Like Santa's list, Johnny's upcoming innovations were infinite.

Johnny had reached the pinnacle of success. He was now free to grow a paunch and openly talk to himself in the hallways of Acme without ridicule. He could now command a board meeting with a mysterious, fleeting smile. His mind was free to roam through untraveled dimensions of new possibilities, catching glimmers of holographic handsets, luminescent luggage, neon-powered blenders, singing coffeemakers, rapping BBQ grills, phosphorescent computer housings, silent lawn mowers, and a refrigerator that would tie your shoelaces. Johnny's mind was a perpetually spinning merry-go-round of product innovation. Together with his childhood friends, Inspiration, Wackiness, and Reason, Johnny was now free to produce a technological bouquet of dynamic improvements for chemical pumps, hair dryers, medical devices, and to-die-for aftermarket gadgets, all designed to create joy, hope, ease, and bliss for people he would never even meet.

## ⠿ What's a Product Developer to Do?

"Go to China and fail" has never been the mandate, but the last 10 years have created those kinds of war stories. The current state of product development has forced product development companies to require that their parts be produced in China. Therefore, management teams are chartered to solve these numerous offshore problems even though they don't fully know the risks or how to handle them.

In today's US economy, if your competition is manufacturing products in China, you only have two choices as a product developer: manufacture your parts in China or get out of that business. When your competitor is getting a 20 to 40% savings on production, they can be very dominant in your market.

## *What Not To Do*

Product developers often use US-based manufacturing representatives of China manufacturers, or they go directly to China in hopes of finding qualified sources to produce their projects. Typically, these approaches fail in providing qualified parts for a company, and they render no substantial savings due to the extensive learning curve and anomalies of doing business in China.

The resources for the US product developer in China are vast. Thousands of companies offer to solve your manufacturing woes in China. Typically, four types of companies attempt to serve this market: direct manufacturing sources, manufacturing representatives, sourcing companies, and adaptive US manufacturers.

The Hybrid Manufacturing Solution, however, is the only approach that works consistently for the good of everyone involved.

### • Direct Manufacturing Sources

The direct manufacturing source is the Asian manufacturer, typically from Taiwan, that realizes the potential of the US market and develops a strategy to attack it. These manufacturers typically own many factories in China and have the capability to produce products for their customers. The Asian manufacturer sets up a US-based sales and support office and begins selling aggressively to the US market with skilled US sales people. However, the management of the office typically remains in Asia.

The key advantages of using direct manufacturing sources are that your project will get strong attention in the US and China, with the right managerial exposure. Also, your pricing should be very competitive, depending on the pricing strategies and management in China and the US. However, there are many hidden disadvantages in using the direct manufacturing strategy.

While direct manufacturing sounds like a great idea, there are still significant information, culture, and marketing gaps that affect decision making between the Asian-managed sales office, the Asian manufacturer, and the managers in the US. Major decisions on how to build the operation are based on an Asian culture, which seems to be familial, understated, and reserved—the opposite of US culture which is more outgoing, performance-driven, and somewhat open. Typically speaking, US offices managed by Asians that have relocated in the US do not have an understanding of American customers. Additionally, the sales team will lack access to key product development opportunities because they have an inefficient approach to accessing, selling, and servicing customers in the US.

The situation that can arise is analogous to a US business, such as Waffle House, opening up in Beijing on the mere assumption that fast-food waffles are universally desired at breakfast. It is very dangerous to charge ahead when the management team knows nothing of the language, people, culture, or expectations of the market. This lack of marketing intelligence and cultural savvy can quickly cause a business to fail. The internal "personality clash" of different cultural values makes for a non-unified corporate dynamic, which can ultimately harm your product. Of course, you want to make sure that your business is not operating on assumptions and that your product is protected from this destructive vortex.

Typically with direct manufacturing, the US office will focus only on the customer's mass production opportunities to get the largest projects possible, representing the highest risk for the product developer. These crucial projects must get to market on time and within budget, with no margin for error. Normally, these big deals are possible only because of an established, mature relationship built on a foundation of trust. Since communication is based on human interaction and is therefore subjective and somewhat fragile, these relationships require long-term tenure to gain customer assurance. Due to this major investment in the customer relationship, the manufacturer's ability to develop enough relationships to scale the business can become impractical. The typical scenario is that the office will continue to grow very slowly or not at all. Over time, the manufacturer's resources are reduced to cut costs, which directly limits service to the customer, until finally the operation comes to a halt.

If you want to use a direct manufacturer, you should try to find the companies that have a history of using the most effective people in the US and China, personnel who can make things happen—people who still believe they have a great business opportunity. If they have little history in the US, then they will lack the resources your project needs to succeed. If they have been too long in the US, then they may be starting to doubt their ability to succeed, and will begin an unintended withdrawal from the market. Of course, there are some exceptions. There are a couple of firms that successfully execute this direct manufacturing strategy. Their success is attributable to their wisdom to seek and authorize US management to run a US operation, backed up with a lot of cash to invest until it works.

## • Manufacturing Representatives

Manufacturing representatives have been the primary sales strategy for US manufacturing for decades. The use of a "sales rep firm" to gain new customers in new markets was a very efficient approach to growth for a manufacturing company before the world "became so small" with internet communications. However, many offshore manufacturers still use this approach to gain new customers in the US. They don't realize the limitations of this strategy, assume the risk is low, or don't know of any alternatives.

The role of the sales rep is to access his or her network of potential customers and sell the services of the manufacturers they represent. When the sales rep is successful, the manufacturer will pay a commission on the sale while the sales rep continues on to the next sell. Meanwhile, the manufacturer attempts to serve the customer. While this approach worked well for US manufacturers using US reps, it has many issues when the manufacturer is on the other side of the world, in a foreign country with a totally different culture.

**Having a US-based project coordination team on-site is critical to the success of projects in China.** An on-site team is so critical that the main issue with the manufacturing rep model is the China manufacturer's difficulty in understanding the needs of the customer's project. A US-based project coordination team must be physically present to manage the project information and communicate requirements and scheduling to all parties. In the past, manufacturing reps for US companies were not required to handle this role because it was relatively easy for a US customer to communicate in English to a US manufacturer for the production of their parts, even though the relationship was started by the sales rep. Once the communication stream changes from English to Mandarin, things can get complicated really fast.

Other issues stem from the sales reps' lack of real understanding about the capabilities of the manufacturers they represent, since they do not typically see a project through to completion. Reps can also waste a lot of time in the quoting process due to communication challenges. With very limited, if any, control at the manufacturer level, sales reps can't always deliver the promises they make to the customer.

The need for the manufacturing rep to keep working his or her network to sell more inhibits his ability to be involved enough in the customer relationship to scale the operation. The sales rep must sell to earn his income, yet he must invest time in the customer to gain the trust necessary to get production orders. Therefore, the rep's opportunities are finite. Time-intensive relationship building is one reason most rep firms maintain fewer than 10 people in a small office, handling only a few customers.

Product developers should be cautious about putting their faith in this very dangerous model. The charismatic sales people involved come across as trustworthy, but they will still lack the infrastructure and systems required to successfully communicate your project to the China manufacturer. If you choose to use this resource anyway, then be sure to include in your plan to your boss the critical element of your strategy...*hope*. Manufacturing reps know how to play golf and entertain, but hard work, not golf, is the business hero's way.

## • Sourcing Companies

Sourcing companies are typically US-based companies that have experience with Chinese manufacturing. These companies developed a business model to leverage their experience by managing manufacturing programs for companies that desire to move their manufacturing to China or want to test China manufacturing with limited investment.

The product development company will engage sourcing companies with a product or family of products that they need to have manufactured at a lower cost. The sourcing company will evaluate the project and determine the best sources of manufacturing in China, then coordinate the manufacturing of the product. The sourcing company generates its revenue from fees for its services or from mark-ups on the product, depending how best to maximize their revenue. Typically, sourcing companies do not work with a variety of new products from product developers, but focus on transferring existing products to manufacturing sources in China.

Most sourcing companies are started by engineers or buyers who worked for a large product developer using China for their needs. These personnel have experience in China, relations with some factories, and are ready to be entrepreneurs. However, these companies lack any unique competitive advantage other than they were the first to arrive and set up shop. Unfortunately, the China challenge is greater than most sourcing companies had imagined because they underestimated the power of the resources they needed to be successful. The "hidden hand" of infrastructure and overhead was not taken into account. However, hundreds of companies place orders with these firms because they offer a great price or a personal relationship. Typically, the outsourced projects start fine with excess supervision but conflicts are imminent as soon as the next project arrives and the hours of the day are exhausted. One definite advantage of working with a sourcing company is an introduction to proven factories in China with which you can develop your own relationship for future projects.

Sourcing companies lack access to many, well-qualified manufacturing projects since sourcing companies typically do not enter the product development cycle prior to mass manufacturing. Sourcing companies also lack experience and technology in managing a high number of production programs simultaneously, thus inhibiting their growth potential. Many of these companies limit their profitability because they do not develop systems or processes that would apply economies of scale to their projects.

**Do not assume that because a factory is "proven" that it will work like your US suppliers.** There is still a communication and cultural gap that must be bridged accordingly to be successful. It's not as easy as sending off your CAD file and waiting for your ship to come in. Remember these obstacles when dealing with China: Communication and language barriers, technical translations, time zone differences, cultural barriers, different holiday schedules, too many choices, and variables of quality. One final note of caution: **In China, things are never as they seem.**

### • Adaptive US Manufacturers

The influence of competition over the past 20 years has shifted the mindset of many US manufacturers and tool makers who finally realize that China is not going away. If manufacturers want to survive they must find new ways to compete. A current trend that will continue is the transformation of the US manufacturer to become an adaptive global manufacturer—one that imports, in a narrow sense, isolated, specific products from China and abroad. In other words, the adaptive companies determine what is best for them to manufacture in the US, and then have the rest manufactured offshore. By using this approach, the manufacturer can still provide the impression to its customer that their company is a US manufacturer, which is partly accurate. Most importantly, an adaptive manufacturing company can provide the cost-saving services that today's customers demand. Adaptive manufacturing and tool making demonstrate American ingenuity. The goal is to not only survive but to thrive in times of adversity.

The trend toward adaptive manufacturing will continue if US manufacturers plan to stay in business. Those that are successful in executing this strategy will be able to take advantage of the strengths of being a US manufacturer, while having the cost advantage of buying products in China. While this is a very sound approach for the US manufacturer, the new business model that includes both internal and external manufacturing presents challenges.

Under this strategy, adaptive manufacturers must now manage two unique business strategies in a combined business model, which can cause strategic conflicts and confuse the customer. For some parts of the project, the manufacturer can be excellent, but for others, it may be less than optimal. Additionally, their lack of experience in managing external manufacturers, especially those very different from their own environment, will create problems. Adaptive manufacturers also have difficulty in shifting the company mindset to support the non-manufacturing aspect of the business. If their skills were honed in their own manufacturing facilities, they are used to having total familiarity and control over every aspect of the project. When they don't have that control, project managers require a different skill set for success. Typically, the adaptive manufacturer resists the new business model that requires a reduction or change in human resources—the need for fewer and different people than the company currently employs. Manufacturers are typically not known for stellar customer service, so when they enter a service segment of the market, their outmoded or poor customer service techniques have a tendency to rub the customer the wrong way.

The US tool maker is more prone to adaptive manufacturing than the manufacturer, since tooling is done in isolation: a single tool is made, repackaged, and sold—a very simple series of transactions. Currently, many successful US tool makers are already adaptive. They make some tools themselves and import others from China. These tool makers break down and rebuild tools from China to ensure quality. By doing this, they can also claim that the tool was "Made in the US." This method actually works and should really not matter to the customer as long as the tool is what they expected.

Typically US tool makers are the cream of manufacturing labor, and as such have a "prima donna" attitude, meaning they are typically in demand, have more work than they can handle, charge top dollar, and are very slow to respond to a customer. Compared to China, US toolmakers require a long lead time to build a tool; however, it is the highest quality in the world. With the economic "red" scare to the North American manufacturing sector, the number of qualified US toolmakers is rapidly declining so that the current supply is low and demand is high, at least for the time being.

### The Hybrid Manufacturing Solution—The Strategy That Works

In the past, tooling and parts were always made by default in the same location by the same manufacturer. The Hybrid Manufacturing Solution separates tooling and parts to give the product developer the greatest value available for each requirement of his project. The Hybrid Manufacturing Solution takes the adaptive manufacturing strategy to the next level to create more product freedom, flexibility, control, and productivity, while leveraging resource strength and reducing risk. The Hybrid Manufacturing Solution of widening manufacturing options is gaining acceptance as evidenced by a number of research applications in various countries.

Everyone gets excited when they hear that tool making in China means a cost savings of 60 to 80%. What they don't hear is that once material costs and logistics are added into the risk of China, it makes more sense to come back to the US to run production parts.

**QuickTip:** "Hybrid" is derived from the Latin hybrida meaning "mongrel." It refers to something of a mixed origin or composition.

Using the Hybrid Manufacturing Solution, you are executing a "triple win" strategy for the US economy, for China and other competing countries, and most importantly, for you the product developer. US product developers can now go back to developing and manufacturing products the way they are accustomed to in the US, while capturing the powerful savings offered by a global manufacturing engine. Unveiling this "production strategy of the century" actually helps the US manufacturer during changing times and will be the key to a thriving economy in the future. What exactly is the Hybrid Manufacturing Solution and how does it leverage China's amazing tool making capacity against the still strong manufacturing capacity of the US?

### • Tools Made in China, Parts Made in the US

**Make your tools in China; run your parts in the US.** The Hybrid Manufacturing Solution is that simple, but intelligently accessing China is not at all simple. A powerful strategy that all product developers can leverage—and most of the big boys already do—is to use a combination of all the best sources. You should leverage the strengths of both China and the US to get your parts manufactured as efficiently as possible. The objective of the Hybrid Manufacturing Solution is to offer quality production using a truly international manufacturing strategy.

The US tool maker or mold maker is a high-expertise professional, comprising five percent of the manufacturing labor pool in the US, thus a low labor content. The tool maker works in isolation, designing machining tools for injection molding. When you use a tool maker, you are paying for expertise and excellence. The Hybrid Manufacturing Solution is based on sending this five percent of the manufacturing labor to China, which does not impact the US economy severely. The US economy can lose tool maker work to China without going belly-up.

The high-labor content resides with the less-skilled operator down-line from the tool maker, who represents 95% of the manufacturing labor content. The operator is the person who actually runs the press to make the plastic parts. If this huge percentage of work goes to China, it decimates the US economy. Therefore, the Hybrid Manufacturing Solution involves sending only the tool maker work to China for the five percent of labor in tool design, but keeps the 95% operator work at factories in North America.

The result of the Hybrid Manufacturing Solution is that product developers save money on manufacturing so that they can develop more products, which drives more manufacturing and gives consumers more options. It also allows the US manufacturing labor base to grow in spite of the downward trend. Companies have the opportunity to make excellent use of existing talent that is not being used by big companies going to China, an essential reassignment of labor that will maintain a healthy manufacturing sector in the US.

### • China Today, Timbuktu Tomorrow

Following a hybrid product development strategy that uses China and the US, other countries are quickly shape-shifting to capture new manufacturing opportunities. You can now prepare to better leverage parts of this widening world puzzle to make your product development faster and more economical. While every project is unique and requires its own assessment, this strategy is very useful for product developing in the low to mid-volumes (1,000 to 500,000 units). As your requirement moves closer to the 500,000 level, then restrictions and risks start to shift and intensify. Serious consideration has to be given regarding the true nature of the part or product, including intellectual property, materials, tolerances, fit, timelines, and control because the magnitude of risk is much higher.

> **QuickTip:** Don't be like the guy who spends more than he saves by driving out of his way for cheaper gas.

## • How Do I Save Time and Money in China?

Using the Hybrid Manufacturing Solution, product developers can realize a 25% cost savings on logistics, save an additional 20 to 40% on the tool making, and avoid the excessive risk of manufacturing parts in China. With the Hybrid Manufacturing Solution, you get product on the shelves in a quick, reliable way. Here are several more ways to save time and money in China.

When purchasing production tooling made in China from a US-based provider, plan months ahead. Always have a padded production schedule of eight to twelve weeks to avoid pushing your China supplier into rush mode and associated errors. Since shipping your tool back to the US by plane costs five times more than ocean shipping, build two to three weeks into your schedule for ocean shipping.

Typically, US companies will send an engineer to a factory in China for the first two weeks and last two weeks of a project. This is a huge mistake as they miss out on guiding everything that happens between visits. Make sure to place engineers at the factory during the entire mold making process. While this may seem impractical at first, it ultimately saves you money because you have control over all of the quality decisions which results in a reduction of rework.

Informed customers always ask for and inspect trial parts before accepting the tooling. Make sure that you see and feel the parts before they are packed for shipping. If the parts do not meet your standards, the Chinese manufacturer is usually happy to fix the tool it made. However, it is important to realize that even good trial parts do not mean that you have a good tool. Under pressure to make acceptable sample parts, the Chinese manufacturer is often using unreasonable process parameters to get parts that you will accept. These parameters can include very long cycle times, which will increase your part price significantly, or use high pressures or temperatures, which can degrade the material. With these process parameters, your parts may appear to be from a good tool, but they are not. The only way to combat this challenge is to have your own qualified engineers at the factory helping process the parts and making sure the tool is really good.

Know your vendors and clearly communicate what you want. Have a complete plan for the project before starting, outlining all tasks and action items. Allocate plenty of time for a good mold design—approximately 10 weeks, plus 3 weeks for ocean shipping. Use a web-based Part Management System to coordinate all project updates. Always respond to the program manager's calls and emails immediately to make a positive impact on decisions at hand.

## • How Do I Waste Time and Money in China?

Beware of snake oil and chicanery. There are a zillion non-qualified brokers who have nothing to sell you in China for a very good price. These purely intermediary people offer no added value, no technical expertise, no networking, no marketing intelligence, and not even engineers at the factory. It's easy to get burned by paying the wrong people to get things done.

Because suppliers in China believe and say they can do everything, it's easy to select a poor supplier based on limited information. Make sure that your plastic molded part is designed for manufacturability (DFM). The formal DFM process checks the part's draft, parting lines, and features to verify that it is indeed "injection moldable." Knowing the physics of your materials and validating them will also save you. You'd be surprised how many companies invest in a tool and never get a part made because they did not accurately anticipate how their geometries would interact with selected materials. Some materials have a six to eight week back log, so be sure to order material (especially special materials) with your tooling.

Prevent poor data translation from CAD to IGES with data validation, matching data to drawings. Lost features create missing holes in your design, and this always means backtracking. Last minutes changes to the design, such as adding a rib, are also time-wasters. Working with the wrong revision of CAD data happens frequently because parts get revised dozen of times, files change hands at midnight, and employees don't always look before they leap.

Finally, there are many common time-wasting scenarios that can ruin your project and cost your company. Doing business in China, it's easy to underestimate the difficulties of translating "technical English," shipping time and costs, the importance of very regular verbal updates and actions, and the effect of tariffs and duties. Overloading a supplier with too many projects can also waste time because you don't know their true capacity. Make sure you understand the limitations of a supplier's systems to handle projects.

Don't wrongly assume that visiting the factory during kickoff will make things work out. Also tweaking tools in the US can be challenging. Unfortunately, you cannot ship tools back to China for tweaking or repair due to prohibitive shipping costs and China duties for importing. Always check to make sure the tooling was not welded in China. Realizing these things too late creates problems in texturing back in the US.

### Little China's Everywhere

The future of manufacturing is wide open. The world will continue to get smaller and smaller as more countries develop their infrastructure to be the next "manufacturing capital" of the world. The manufacturing world watches as Vietnam gears up for global production, and opportunities expand into Eastern Europe, where in the Ukraine, for example, Cold War talents that made weapons are now making peacetime pots and pans.

Along with an increasing use of China manufacturing, US product developers will also expand their use of other non-domestic manufacturing resources. Other regions that will increase manufacturing custom parts and products include Korea, Vietnam, India, and South America. Regions offering an option to China will become very important politically and economically to help offset the increasing dominance generated by China and its manufacturing base. Therefore, US companies need to access these other regions now, either directly or indirectly, in order to minimize the risks associated with China, while leveraging additional economic benefits.

Adapting to this swirling global dance floor is the new way of manufacturing. We must understand the challenges of changing resources, environments, and politics. In global manufacturing, there are many hurdles to overcome when you no longer have a contiguous legal system to protect you, a common language that is understood by all, or a unified sense of expectation, thanks to the diversity of all the global players—challenging indeed, but only the strong accept the choice to change, to thrive.

It's time to pack your bags. You are now ready for China. Here's your passport, your ticket, and the keys to a brand new hybrid car. If you're not a traveling DIYer (Do-It-Yourselfer), then **use a Hybrid Manufacturing Solution expert.** It will save you not only a long plane ride but also your time, money, hair, marriage, and waistline.

## ▓ The Keys to a Brand New Hybrid Smart Car

By now you've got a more realistic view of how to access China. You understand its strengths and weaknesses as the manufacturing capital of the world. You know that the Hybrid Manufacturing Solution is a proven strategy that was born from a desire to help the US economy and the product development customer. You understand the relationship of production to the earlier additive fabrication processes used to verify design. You get that product development is like gestation and production is like giving birth. You've discovered that doing business in China is a wonder, a mystery, and a pain. Quite possibly, you are ready to take the next step across this tiny world to realize amazing benefits in production tooling while China reigns.

As Johnny would say,

**"Git 'er done!"**

## Chapter 8

# Business on Purpose

### A GARDEN OF GOOD AND GOLD

*"It's better to have tried and failed
than to only dream and wonder!"*

## :: Quickparts—A Personal Passion

Quickparts provides custom manufacturing services to engineers and designers looking to produce plastic and metal parts from 3D CAD files. Quickparts is North America's largest provider of custom-designed parts, from rapid prototyping (RP) to production parts. The company began as a spinoff from a traditional engineering services business that worked hard to help companies develop products faster. This experience helped us further understand the challenges that today's product developers have when dealing with manufacturers to get custom parts made quickly and economically. After years of not being treated as a valued customer by manufacturers and waiting days for information, we realized that there had to be a better way to serve this market. In 1999 the Internet was red hot, and we had special technology for manufacturing process analysis sitting on the shelf just waiting for a problem, and a few young, bright, dedicated guys who wanted to change the world…this was the beginning of Quickparts.com, Inc.

For most of my career, I have had an obsession with efficiency and optimal performance. This has extended to most, if not all, parts of my life (just ask my wife—she loves the chats about how to optimize the performance of our relationship). Being an engineer by degree, I started in product development. When I began, the product I was developing was the International Space Station, this "little" project that is the basis of universal exploration and finding the meaning of life! While this project was amazingly inefficient, it used the latest technologies available in the process. In 1990, the technologies of "tomorrow" included a complete solid model in CAD (at the low cost of $250,000 per seat), Stereolithography, Finite Element Modeling and Analysis, and super-computer-level software programs.

In my small world, one thing that was driving the inefficiencies was that we did not verify any design during the design process. In other words, we did not buy any insurance that our designs were right, reasonable, manufacturable, or economical. The only thing that mattered then was if the part was maintaining a strict diet of low-weight aluminum and titanium and that the chosen materials were "on the list." Otherwise, we would release the design to manufacturing and wait for our calls to be ridiculed by manufacturing, wasting years—yes, years—by having to start the process over from the beginning.

As mentioned earlier, at that time there was a distinct wall between engineering and manufacturing. If you were too young and naïve and climbed over the wall to say, "Hey, why don't we work together earlier in the design?" you would be laughed at out loud and promptly tossed back over the wall. Ask this enough times, and you get to spend some time with your manager (which was rare and not very positive). Since all experiences in life have a purpose, it was these early experiences that defined my career path to break down that wall between engineering and manufacturing and drive greater efficiency in developing new products.

As it turned out, my obsession with efficiency coincided with the advent of the greatest tools ever provided for product development. These tools included easy-to-use and affordable solid modeling with CAD, which eliminated the dependency on interpreting the language of 2D drawings and made the design job more "real." Finite Element Analysis (FEA), materials libraries, rapid prototyping, and the Internet all combined to make a paradigm shift in the thinking about how long it should take to make a production part.

I soon learned that by combining all of these processes into one product development "toolbox," companies would be able to produce many more products in less time. Together, these tools make it possible to get parts made quickly, thus our company, Quickparts.com, was born. The processes associated with Quickparts include the primary technologies of rapid prototyping and efficient manufacturing to produce parts faster. These processes reduce the time of the past from weeks to make a model, to hours to produce the actual part that can be used in the product. These processes radically reduce the traditional 12-week schedule for making a production-level injection mold to only two to four weeks. What was once considered impossible is now the expected and required to further drive the product development process.

## ⠿ The Quickparts Model

Quickparts was designed to be the "Home Depot" of custom-designed parts. If you need a hammer, you go to Home Depot because they have many hammers, excellent customer service, great prices, and buying is easy. At Quickparts, we do the same thing for the customer that needs custom-designed parts. We make it very easy to buy with online instant quoting, and we offer all the major manufacturing processes, provide outstanding and customer-oriented service, along with competitive pricing. We deliver everything from one-of-a-kind "onesies" using rapid prototyping to hundreds of thousands of parts made with injection molding tooling.

With Quickparts in the market, large and small companies no longer hassle with the manufacturer; they can now focus their valuable time on product development. True providers like Quickparts have become the partner to these companies because we have the ability to perform and make it easy for them to do business using a global strategy.

## ▪▪ The Purpose of Business—The Expanding Spiral

While "parts are the center of the universe," the business entity is the most elemental unit in a free enterprise system. Fueled by the individual and launched socially, business is the glue that holds the globe together. If it were not for business, we wouldn't be at all concerned about China. When the nightly news anchor announces, "Declining US dollar"...we cringe. No one likes the sound of that. It hurts our soul and sounds alien to the wholesome, booming American image we have all internalized. So we have to ask ourselves two central questions: What are we doing as a business nation and what is the purpose of business *really*?

> **QuickTip:** Innovation is the differentiator in a free enterprise. It is not easily "owned" or replicated.

Did you fall asleep at your desk 10 years ago only to wake up on a much smaller planet? Did you really think things wouldn't change?

The purpose of business is essentially to satisfy the customer, develop people, forge leaders, grow communities, and by doing so, provide a positive return to shareholders. All of the players are interconnected in an ever-expanding spiral.

### Satisfy the Customer—Fruit and Labor

If you are in business, your entire purpose is to serve a customer, even if that "customer" is a co-worker or even your boss. If you work for a large, global company, it can be very hard to understand who your customer really is and how your role impacts them. Internal or external customers, everyone's role is to satisfy or you would not have a job. The purpose of business is to satisfy the needs of customers. This is the main purpose and without it, there is no reason for the business to exist. **Does every business you engage serve your needs?**

One of the many great things about the US is that we live in a system of capitalism where customers vote with their dollars. Demand service from every business. This is the only way to drive a company to improve, and, of course, drive their efficiencies to optimal performance. The cashier at FedEx Kinko's, Starbucks, or McDonald's may not like you, but in the next hour he or she will do better, I guarantee it.

### Develop People—Leaders will Blossom

In the world according to Ron, the next purpose of the company is to develop people and to forge new leaders. As a leader, foster an environment in which people can learn, grow, and expand their awareness. Employees, team members, associates go by many names, but they are always people. So, it is imperative that every business is committed to developing its people to be the best they can be. Define goals clearly, and demand excellence of everyone. By doing this, the company is then able to incubate individuals to get the most out of their work and their lives. The employee relationship is not a trivial one. As a leader, you provide the environment in which an individual finds his or her unique potential, found in their innate passion. If you first fit the person to their passion, then fit the passion to their job, they will be passionate about their job and passionate about serving. As a true leader, you assist your people to wake up and pursue their passion—if it means helping someone "let go" of their job to pursue their passion, do it!

What would you do for free? That million-dollar question is a quick "passion pursuit" test. If what you are doing is it, then great! If not, then challenge yourself because **you must have passion in what you are doing to be the best and to be fulfilled.** This is easy to see every fall Saturday when college football players give 110% of their all to be the best. By developing people to be their best, grow in their lives, and gain confidence in their capabilities, as a leader you are serving the next purpose of business: people potential. You want to surround yourself with people who are passionate about reaching their own potential.

> **QuickTip:** Use unbridled market forces to design a better world.

Do this through setting clear expectations, acknowledging good results, and being candid about performance. Avoid negative criticism that kills the soul. It has no place in any kind of relationship, period.

### Grow Community—It's Great to Cross-Pollinate

"Think globally, act locally" is a wonderful business philosophy. No business can exist without a community that is "host" of the business, providing support that is required for growth and efficiency. By forging leaders in the business, you can sleep well knowing that these folks go home each night to develop more leaders from family and friends. They will make a positive social contribution as confident leaders in their schools, churches, neighborhood associations, and softball teams. As an inspired person, your employee can drive the individuals and groups in their greater network to grow and develop to be the best they can be. As humans encompassed within Maslow's hierarchy of needs, we all naturally want to rise be the best we can be and serve at our highest capacity. Local communities impact the global network. As we said at the start: **Everything matters...even the butterfly.**

### Return to Investors—Gardeners of the World

The last purpose of business is to provide an appropriate return to the investors of the business. These "gardeners of the world" risk their funds to seed businesses with monies to develop new products or services. Without shareholders and their "seeds, water, and fertilizer," businesses could not exist and the world would suffer a great loss. Can you imagine living without your Starbucks double espresso, non-fat mocha latté? Next time you sip a hot one, you might remember that investors made your awesome "brand" experience possible. Or try walking to work while you say a prayer of thanks for the investors who helped develop your new car on their dime.

## ⠿ A Fond Farewell

We've covered a big story about an emerging paradigm in manufacturing. We've looked at product development through a microscope and telescope, relating the microcosm of manufacturing processes to the macrocosm of world superpowers, in search of simple business truths. We've related the relationship of parts, products, businesses, societies, and countries to the greater cosmos and the eternal spirit of creativity that resides in all of us. We've looked at manufacturing through the eyes of a business leader, an engineer, a mystic, and a passionate patriot. We've looked through the eyes of a fairytale hero, Johnny Quickparts, who, we are hoping, is a bit like you.

### A Very Tall Tale—Johnny Quickparts Takes a Rorschach Test

After Johnny's promotion, Acme's invisible owner decided to shake things up a little more. He sent in an independent executive coach to thoroughly clean house. The coach administered only one very simple Rorschach inkblot test: the image looked like a large V to most human beings. The coach assured everyone that there was no right answer! He asked them to free associate and write down the first V-word that came to mind.

After testing 100 top managers, the coach delivered the following results: 85% of those tested said the "V" stood for Victim. 6% thought of Vicks VapoRub, 3% thought of

Vermin, 2% thought of Volleyball, 2% thought of Vixens, and 1% said Vitamins. Only Johnny saw "Victory, the fabulous kind created by kings who care about people in a golden way." By the end of the day, only the Vitamins, Volleyballs, and Vixens had jobs. Johnny the Victorious was made CVO, *Chief Vision Officer.*

As Johnny would say,

**"Think it, feel it, do it!"**

# QuickSMART Glossary

**ABS** – see Acrylonitrile Butadiene Styrene

**Acrylonitrile Butadiene Styrene** – a common thermoplastic used to make light, rigid, molded parts. Also, a material used for the FDM process.

**Additive Fabrication** – a manufacturing process in which a part is made by virtually slicing a part into layers; producing each layer; and the subsequently adding or joining the layers together in the appropriate order to produce a complete part. Also referred to as Rapid Prototyping.

**AF** – see Additive Fabrication

**ASME** – American Society of Mechanical Engineers

**Bas-relief** – pronounced: bah-releef; a subtractive method of sculpting that carves away select surface areas of a flat piece of stone or metal, resulting in a raised pattern, often used in architectural detail.

**CAD** – see Computer-Aided Design

**CAM** – see Computer-Aided Manufacturing

**Cast Urethane** – a formative manufacturing process in which durable, real-looking parts are made from a two-part polyurethane material and formed in a silicone mold under relatively low pressure. This process is ideal to make up to 50 parts that are plastic and simulate some thermoplastic characteristics. The process has three distinct steps: pattern, RTV tool, and cast part.

**CMM** – see Coordinate Measurement Machine

**CNC** – see Computer Numerically Controlled

**Computer-Aided Design** –sophisticated software in which the physical world (parts, buildings, etc.) can be represented for design, analysis and verification. CAD data is the input for RP, thus a requirement to take advantage of the technologies.

**Computer-Aided Manufacturing** –sophisticated software used to convert data, typically from CAD, into useful data to easily program and control manufacturing processes. CAM is essential to CNC complex or organic parts.

**Coordinate Measurement Machine** – a device for dimensional measuring. It is a mechanical system designed to move a measuring probe to determine the coordinates of points on the surface of a part.

**Computer Numerically Controlled** – the use of a special processing language that controls the movements and actions of a manufacturing system. Typically used for machining operations in which the program provides instructions to the mills or lathes in machining a part.

**CU** – see Cast Urethane

**DDM** – see Direct Digital Manufacturing

**Design for Manufacturability** – an approach to design in which considerations are made during the design process to make a part easy to manufacture. By applying the principles of DFM, manufacturing issues and defects are reduced and the cost of the parts is less.

**DFM** – see Design for Manufacturability

**Direct Digital Manufacturing** – the latest term used to describe the production of end-use parts from an additive fabrication system. SME has endorsed this term and define it as the process of going directly from an electronic digital representation of a part to the final product via additive manufacturing.

**ECO** – see Engineering Change Order

**EDM** – see Electrical Discharge Machines

**Electrical Discharge Machines** – uses electrical charge to burn away excess, unwanted material

**Engineering Change Order** – also known as engineering change notices (ECN) or just engineering changes (EC). They are part of a design system to allow for and control changes that are required to the design or manufacturing of a part. Engineering changes can be very disruptive to the product development process and should be reduced as early in the process as possible with design verification and product validation.

**FDM** – see Fused Deposition Modeling

**FEA** – see Finite Element Analysis

**Finite Element Analysis** – a computer simulation system that virtually assesses the behavior of a part to better predict how it will perform in the physical world. With FEA, a design can be modified to reduce the likelihood of failures.

**Fused Deposition Modeling** – an additive fabrication process that creates parts by extruding plastic wire through a CNC-controlled extrusion nozzle, layer by layer. FDM produces some of the strongest and most durable parts of the leading AF technologies.

**Grok** – rhymes with "walk"; a fun, friendly science fiction verb, coined by author Robert Heinlein which means "to understand something so well that it is fully absorbed into oneself."

**Low-Volume Injection Molding** – a manufacturing method that creates injection molds or tools to produce functional parts from thermoplastics in short runs of up to typically 50,000 parts. LVIM offers the similar quality and accuracy as production tooling, however it is faster and typically more economical. LVIM is an excellent tool for products that have low production requirements, such as a medical application; or parts are needed quickly while a production tool is being produced; or to verify a part design can be manufactured with injection molding.

**Low-Volume Layered Manufacturing** – a very powerful, evolving trend in which end-use parts are manufactured from additive fabrication systems. Also known as DDM or rapid manufacturing, this approach provides more flexibility in the design process by eliminating the typical rules of DFM. The designer can combine more functions into a part or change their design as needed after the part is released.

LVIM – see Low-Volume Injection Molding

LVLM – see Low-Volume Layered Manufacturing

Product Developer's Toolbox –a choice selection of CAD-friendly product development tools that bring about product innovation faster and better. It is defined as the application of technology and processes for the manufacturing of functional parts quickly and economically.

Rapid Prototyping – the early name given to a class of automated machine technology that quickly fabricates physical 3D parts from electronic 3D data by building the part in layers. Some of the leading RP technologies include Stereolithography, Fused Deposition Modeling, and Selective Laser Sintering. There are dozens of competing technologies that can produce physical parts by additive fabrication.

Room Temperature Vulcanization – the process of making a rubber material (vulcanization) by mixing a compound of silicone materials at room temperature. RTV molds are made from an impression of a master pattern in the rubbery material, leaving a negative space to fill. The RTV mold is used to cast parts from a urethane material.

RP – see Rapid Prototyping

RTV – see Room Temperature Vulcanization

Selective Laser Sintering – an additive fabrication process that creates parts by fusing or sintering particles of powdered material with a hot $CO_2$ laser, layer by layer. SLS produces very durable parts that can withstand high temperatures. Also, the SLS process is the most efficient in that it does not require support structures and can build parts in a 3-dimensional build envelope.

**SL** – see Stereolithography

**SLA** – see Stereolithography Apparatus

**SLS** – see Selective Laser Sintering

**SME** – Society of Manufacturing Engineers

**Standard Tessellation Language** – a language that produces the much needed STL file from the electronic CAD design data. The STL file is generated by representing the model with triangles on all surfaces, then outputting the vertices and normals of the triangles into a useful file. The STL file is required for most additive fabrication technologies. It is considered a standard output of the CAD software such as IGES and STEP.

**Stereolithography** – an additive fabrication process that creates parts by tracing the cross-section of a part with a laser onto a vat of photocurable resin. The laser provides the energy required to initiate a phase change of the resin from a liquid to a solid. This process is repeated for each layer until the part is completed. SL was one of the pioneers of the RP industry. It is said the name is derived from stereo, meaning three dimensional and lithography, meaning to print, thus providing the compound word meaning to print in three dimensions.

**Stereolithography Apparatus** – the machine for the Stereolithography process. The term SLA is a common name given to the technology from the early days in the technology, but really was referring to the machine. Some will refer to Stereolithography as SL to be more accurate.

**STL** – see Standard Tessellation Language

# Index

# NOTES

# NOTES

NOTES

## About the Author

**Ronald L. Hollis, Ph.D., P.E.** is the President, CEO, and Co-founder of Quickparts.com, Inc. He provides experience, leadership, and knowledge in the development of a leading business in the product development market. With his vision and drive, he has been able to lead an innovative, entrepreneurial, and fast-growing company that helps product developers succeed.

Ron and his company have received many awards and recognitions including: 2005 and 2006 Deloitte Fast 500; Finalist Small Business Person of the Year from Atlanta Metro Chamber of Commerce and Atlanta Business Chronicle; 2004 Ernst and Young Finalist Entrepreneur of the Year; 2004 Innovator of the Year by Catalyst Magazine; Top 50 Entrepreneurs of Atlanta; Hot 100 List from Entrepreneur Magazine; 2004 INC 500 company.

Ron is a graduate of MIT/INC Birthing of Giants, the former President of the Atlanta Chapter of Young Entrepreneur's Organization (YEO), a member of Young President's Organization (YPO), a corporate sponsor of Hands-On Atlanta charity organization, and serves on the boards of other entrepreneurial companies.

He earned a BS in Mechanical Engineering from the University of Alabama, worked as a design engineer on Space Station Freedom for Boeing, earned an MS in Engineering and a Ph.D. in the management of technical business, and is a registered Professional Engineer. He is also a private pilot and enjoys boating.

Ron currently lives in Atlanta with his wife, Melanie, of more than 15 years and their son, Jackson.

For more information or to order more books,
please visit us at:

*www.betterberunning.com*

\*\*\*\*\*\*\*\*\*\*\*\*\*\*\*\*\*\*\*\*\*\*\*\*\*\*\*\*\*

We are dedicated to continuously improving
everyday. If you find errors, issues or suggestions
to improve, please send us an email at:

*ron@betterberunning.com*

\*\*\*\*\*\*\*\*\*\*\*\*\*\*\*\*\*

Thank you for the opportunity to serve your needs.